Differential Equations - Theory, Modeling, Data Assimilation and Algorithms

Edited by Don Kulasiri

Published in London, United Kingdom

Differential Equations - Theory, Modeling, Data Assimilation and Algorithms
http://dx.doi.org/10.5772/intechopen.1006213
Edited by Don Kulasiri

Contributors
Channa Rajanayaka, Don Kulasiri, Ines Ellouze, Jeevabharathi Ranganathan, Jesús García-Martínez, Jesús García-Ravelo, Jesús Morales, Jing Yang, José Juan Peña, Linda N.A. Botchway, Maryam Ben Salah, Parul Tiwari, Perrin G. Kibiti Pembe, Zaer Salem Abo-Hammour

First published in London, United Kingdom, 2025 by IntechOpen
IntechOpen is the global imprint of INTECHOPEN LIMITED, registered in England and Wales, registration number: 11086078, 167-169 Great Portland Street, London, W1W 5PF, United Kingdom

For EU product safety concerns: IN TECH d.o.o., Prolaz Marije Krucifikse Kozulić 3, 51000 Rijeka, Croatia, info@intechopen.com or visit our website at intechopen.com.

British Library Cataloguing-in-Publication Data
A catalogue record for this book is available from the British Library

Differential Equations - Theory, Modeling, Data Assimilation and Algorithms
Edited by Don Kulasiri
p. cm.
Print ISBN 978-1-83634-950-1
Online ISBN 978-1-83634-949-5
eBook (PDF) ISBN 978-1-83634-951-8

If disposing of this product, please recycle the paper responsibly.

Meet the editor

Professor Don Kulasiri is a Professor (personal chair) and has been Head of the Centre for Advanced Computational Solutions (C-fACS) at Lincoln University since 1999. His research theme is to understand the mathematical basis for biological and environmental phenomena based on physics broadly construed to assimilate data-driven machine learning and AI in phenomenological models. He has been a Visiting Professor at the Mathematical Institute, Oxford University, the UK, since 2008, Princeton University, USA (2004,2006), and the Mechanics and Computation Division, Stanford University, USA (1998) and a New Zealand Centre at Peking University Fellow (2018), and a Fellow of the Modelling and Simulation Society of Australia and New Zealand (MSSANZ). He has published over 215 publications and authored 7 research monographs on his research with leading international publishers.

Contents

Preface

Over the past two centuries, differential equations have been central to the development of science and engineering, providing the foundational framework for quantitative analysis and offering a systematic means of describing how systems change over time. Whether tracking the orbits of celestial bodies, capturing the complex dynamics of biological systems, or modeling fluctuations in financial markets, differential equations serve as essential tools for converting physical laws, empirical data, and conceptual models into forms suitable for rigorous analysis.

Differential Equations – Theory, Modeling, Data Assimilation and Algorithms has grown out of many years of research and interdisciplinary collaboration by the authors, experts in their respective fields. This book reflects the critical role that the integration of theoretical concepts with computational techniques plays in revealing the behaviour of complex systems. It aims to close the gap between the rigor of mathematical foundations and the practical challenges encountered in modeling and inference, providing treatments that are accessible to advanced students while also serving as a valuable resource for researchers and practitioners.

In today's scientific environment, the importance of differential equations cannot be overstated. However, the traditional approach of examining them in isolation from data and computational considerations has become inadequate. With the growing availability of data, there is a pressing need for methods that can align theoretical models with observational data, allowing for the estimation and refinement of model parameters, states, and even structural components. It is within this interface between models and data that data assimilation plays a pivotal role.

Concurrently, advancements in numerical algorithms for solving differential equations have greatly expanded our capacity to study complex systems that defy analytical solutions. Yet, computational tools should not be treated as opaque procedures; a deep understanding of their stability, limitations, and interactions with data is essential to draw meaningful scientific conclusions.

I am thankful to all the authors who contributed, and to Ms. Ivana Barac, Publishing Process Manager and the staff at IntechOpen, whose assistance was crucial in completing this book.

Don Kulasiri
Centre for Advanced Computational Solutions (C-fACS),
Lincoln University,
Christchurch, New Zealand

Chapter 1

An Investigation of the Role of Clockwork Orange (CWO) in Circadian Rhythms of *Drosophila melanogaster*: A Mathematical Modeling Study

Don Kulasiri and Jeevabharathi Ranganathan

Abstract

The daily behavioral cycles exhibited by nearly all living organisms, known as circadian rhythms, are a crucial feature of life shaped by the Earth's rotation and regulated by internal biological clocks. These approximately 24-hour patterns reflect intricate biochemical and physiological processes. The fruit fly, *Drosophila melanogaster*, has emerged as a pivotal model for studying circadian rhythms, with its genetic and molecular underpinnings extensively characterized. Mathematical modeling is a common tool used to dissect these dynamic systems. This study presents three new models of the circadian pathways of *Drosophila melanogaster*, each integrating three separate transcriptional feedback loops: the classic PER-TIM and VRI-PDP1 cycles, as well as the newly identified Clockwork Orange loop. These models investigate three possible dual functions of the CWO protein, hypothesizing that it may simultaneously activate and repress key circadian genes. They combine established molecular insights with new hypotheses derived from both *in vivo* data and CWO's protein sequence using bioinformatics tools. The models employ a probabilistic ordinary differential equation (ODE) grounded in chemical kinetics to describe how transcription factors bind to and dissociate from their targets. Rather than relying extensively on potentially inconsistent *in vitro* measurements, the study focuses on developing conceptual models and testing hypotheses. Available data were primarily utilized to refine parameter estimates and assess model validity. This methodological choice provided the flexibility needed to probe the molecular functions of CWO more thoroughly. Model behavior was evaluated and validated using mutant data, and the resulting simulations offer insights into circadian biology by clarifying the role of CWO at the molecular level.

Keywords: circadian clock, clockwork orange, genetic networks, rhythms, *Drosophila melanogaster*, mRNA, mathematical modeling, ordinary differential equations, oscillations, protein

1. Introduction

Circadian rhythms are biological responses to daily changes in light, temperature, and other environmental factors, influenced by Earth's axial tilt and orbit around the sun. These rhythms involve synchronized cycles of gene, mRNA, and protein activity over approximately 24 hours, leading to physiological changes in organisms. Light and temperature play a crucial role in regulating these rhythms, which are found in a wide range of life forms, from certain prokaryotes and plants to eukaryotes, including mammals. Even in stable environmental conditions, circadian rhythms maintain distinct characteristics, such as temperature-compensated rhythmicity [1]. They align with the day-night cycle, with light and temperature serving as key timekeepers, or 'zeitgebers'. The fruit fly (*Drosophila melanogaster*) is a commonly used model for studying circadian rhythms due to its suitability for genetic research in laboratory settings. (Nomenclature: names of genes are in lower case italics and proteins are in upper case normal characters.)

This chapter presents three novel mathematical models that describe the circadian rhythm in *Drosophila melanogaster*, each incorporating three interlinked transcriptional feedback loops. Two of these regulatory circuits, VRI/PDP1 and PER/TIM, have been extensively characterized and included in earlier models, such as those developed by Xie and Kulasiri [2]. The primary focus here is on the third feedback loop, centred around the recently identified Clockwork Orange (CWO) protein. Each of the three models explores different possibilities for CWO's dual functionality, postulating that it may act both as an activator and a repressor for several core circadian genes, including *per*, *tim*, *vri*, *pdp1*, and *cwo*. By integrating established molecular knowledge with insights from computational bioinformatics analyses, this research aims to refine our understanding of the potential regulatory role played by CWO in the circadian clock, guided by experimental observations.

2. Biological background

The first circadian rhythm gene to be identified in *Drosophila* through ethyl methane sulfonate (EMS) mutagenesis screens was the *period* (*per*) gene [3]. Subsequent mutational analyses and behavioral studies revealed three key *per* alleles: the arrhythmic loss-of-function allele (*per01*), a variant producing a shortened circadian cycle (*pers*), and another that extended the daily rhythm period (*perl*) [3]. Cloning and characterization of the PER protein demonstrated that it contains a domain structurally similar to those found in Single-minded (SIM) and Aryl hydrocarbon Receptor Nuclear Transport (ARNT) proteins [4–6]. This region, later termed the PAS domain, mediates critical protein-protein interactions [7]. Hybridization assays indicated that PER and TIM, which is the product of another clock gene (*tim*), could form a heterodimer through these PAS domains, resulting in the PER/TIM complex [8, 9]. The discovery of an arrhythmic null allele of *tim* (*tim01*) provided definitive evidence for TIM's essential role in sustaining circadian rhythms [10]. While these advances highlighted the importance of *per* and *tim* in maintaining daily rhythmicity, their exact molecular mechanisms remained elusive until further studies clarified their contributions to the circadian feedback system [11].

The *period* (*per*) gene is subject to autoregulatory control, and mutations within *per* disrupt the rhythmic synthesis of its mRNA. Notably, *per* mRNA and PER protein levels oscillate in anti-phase [11], suggesting that PER functions as a transcriptional repressor of its own gene. Similarly, the *timeless* (*tim*) gene and its protein product, TIM, exhibit autoregulatory behavior, with their oscillations driven by corresponding fluctuations in *tim* and *per* mRNA levels [12].

There is an ~70 bp sequence located approximately 500 bp upstream of the transcriptional start site in *per* gene, which contains a canonical E-box motif (CACGTG) essential for transcriptional activation. A similar E-box element was found in the *tim* promoter, also required for its expression [12].

Subsequent mutagenesis studies revealed the essential roles of two genes, *clock* (*clk*) and *cycle* (*cyc*), in maintaining circadian rhythmicity. These findings enhanced understanding of the PER/TIM feedback loop, as *per* and *tim* expression was significantly reduced in *clk* and *cyc* mutants, leading to arrhythmic phenotypes [13]. CLK and CYC were thus proposed to function as positive transcriptional regulators of *per* and *tim*.

Both CYC and CLK proteins harbor PAS domains and basic helix-loop-helix (bHLH) motifs, which are characteristic of E-box-binding transcription factors. It was proposed that CLK-CYC heterodimers activate *per* and *tim* transcription by binding to E-box motifs in their promoters *via* their bHLH domains. Subsequently, PER and TIM proteins interact with CLK and CYC via their PAS domains, displacing them from the E-box and thereby terminating transcription [14].

The temporal expression of *per* and *tim* mRNAs is approximately anti-phase to that of *clk* mRNA [14]. While *clk* transcript levels peak shortly after dawn, *per* and *tim* transcripts reach maximum levels in the early evening. In per^{01} and tim^{01} null mutants, *clk* mRNA levels are markedly reduced compared to wild-type, initially leading to the erroneous conclusion that PER and TIM positively regulate *clk* transcription [15]. However, in clk^{jlk} mutants, which express a nonfunctional CLK protein, *clk* mRNA levels are significantly elevated [16]. Further investigation revealed that the *clk* promoter lacks canonical E-box elements, indicating that it is not a direct target of CLK/CYC regulation [14].

This paradox was further underscored in $per^{01}clk^{Jrk}$ double mutants, where *clk* mRNA levels remained constitutively high [16], suggesting additional regulatory mechanisms. A rhythmically expressed PAR domain-containing protein, VRI (Vrille), was subsequently identified. The *vri* promoter contains an E-box, implicating activation by CLK/CYC. Overexpression of VRI led to a substantial reduction in *clk* mRNA levels, and the discovery of a VRI binding motif (VRI box) within the *clk* promoter supported a direct repressive role for VRI in *clk* transcription [16].

In 2003, PDP1 (PAR Domain Protein 1), another PAR domain factor, was identified based on its homology to VRI [17]. Both *vri* and *pdp1* promoters contain E-box motifs and are regulated by CLK/CYC binding [17, 18]. Sequence analyses revealed that VRI and PDP1 belong to the bZIP (basic leucine zipper) transcription factor family, suggesting they bind to similar DNA motifs. Experimental data confirmed that both proteins interact with a shared region in the *clk* promoter, termed the V/P-box. VRI and PDP1 exert opposing effects on *clk* transcription *via* competitive binding to this site—VRI represses, while PDP1 activates *clk* expression [17, 18].

By 2007, six core transcriptional regulators of the *Drosophila* circadian clock had been identified, excluding the later-discovered Clockwork Orange (CWO). These proteins fall into two functional categories: activators (CLK, CYC, PDP1) and repressors (PER, TIM, VRI, CWO). Collectively, they form a complex transcriptional feedback network that orchestrates circadian rhythms through finely tuned cycles of gene activation and repression.

3. Objective

A component of the circadian clock called Clockwork Orange (CWO) has been discovered in *Drosophila* [19–22]. CWO is a member of the Hairy-Orange domain family of transcription factors, which includes mammalian repressors such as HES-1, HES-2, and HES-3. In pacemaker neurons, CWO protein levels oscillate in a pattern similar to CLK/CYC-activated proteins like PER and TIM, though with slightly lower amplitude [19]. Promoter analysis of the *cwo* gene revealed up to 20 E-box elements within the first intron of the 5′ region that are targets for CLK/CYC binding [20]. This suggests that CLK/CYC likely regulates *cwo* transcription, consistent with data showing that *cwo* mRNA levels fluctuate in sync with other core clock genes—including *vri*, *pdp1*, *tim*, and *per*. Supporting this, *cwo* transcripts decrease in *clkjrk* mutants and increase in *per01* mutants, mirroring the behaviour of other E-box-regulated clock genes [19].

Studies of *cwo* mutants, which produce defective CWO protein under constant darkness (DD) conditions, showed marked disruptions in the oscillations of CLK/CYC-regulated genes. These effects mimic the loss of the VRI/PDP1 feedback loop, resulting in dampened rhythms and lengthened periods of E-box-driven clock gene expression compared to wild-type flies [20]. Such findings highlight the crucial role of CWO in maintaining strong circadian rhythms.

As a basic helix-loop-helix transcription factor, CWO was hypothesized to bind to E-boxes within other clock genes. Immunoprecipitation assays using tagged CWO confirmed its interaction with the E-boxes of *vri*, *pdp1*, *tim*, and *per*, as well as with its own promoter's E-boxes [21]. Transcriptional studies in *Drosophila* S2 cells under light-dark (LD) and constant darkness (DD) conditions demonstrated that CWO participates in a negative feedback loop by repressing the expression of all E-box-containing clock genes, including its own [21].

A null allele of *cwo*, termed *cwoB9*, was produced by EMS mutagenesis and encodes a truncated 36-amino acid protein, in contrast to the full-length 698-residue version. Behavioural studies of *cwoB9* mutants under LD and DD conditions revealed a $\sim$ 2.5-hour extension in activity rhythms. In these mutants, E-box-containing clock genes—*pdp1*, *vri*, *tim*, and *per*—showed reduced rhythmic expression, while *cwo* transcripts themselves remained elevated [22].

Early cell culture experiments suggested that CWO functions broadly as a transcriptional repressor, binding to E-boxes in clock genes to suppress CLK/CYC activation [19, 21]. However, the surprising drop in *per*, *tim*, *vri*, and *pdp1* transcripts in *cwo* mutants also points to a potential activator function for CWO [22].

In summary, these studies highlight the complex and possibly dual functions of CWO in circadian regulation. This ambiguity underlies the main objective of our research: to refine and clarify the functional roles of CWO, laying the groundwork for experimental validation of the hypotheses generated in this work.

4. Development of models

Given the critical role of the CWO component and its ability to assume multiple functions in clock oscillations, we developed three ordinary differential equations (ODEs)–based models—designated Model A, Model B, and Model C—to perform in silico experiments aimed at identifying the most plausible molecular mechanisms underlying CWO's dual functionality. Our hypotheses were informed by existing wet-lab data and bioinformatics analyses of the CWO protein. Accordingly, each model represents a distinct conceptual framework described by a unique set of probability-driven ODEs. These models were solved and parameterized, with results detailed in the following sections. Specifically, Model A posits that CWO binds to E-boxes independently; Model B suggests that CWO forms a heterodimer complex (CWD) with a hypothetical protein (HP) to bind E-boxes; and Model C assumes that CWO binds alone but introduces a novel post-translational complex (CWPT).

Our approach builds upon the conceptual and probabilistic framework established by Xie and Kulasiri [2], extending it to develop Models A, B, and C. A key feature of their model is the incorporation of all primary transcription factors (TFs) identified through wet-lab studies—including CLK, CYC, PDP1, PER, VRI, and TIM. While earlier models omitted certain components, their Drosophila circadian model successfully simulated sustained oscillations for all six proteins and corresponding mRNAs known at that time [2]. The framework relies on a positive transcriptional feedback loop involving *per* and *tim* genes, along with a regulatory feedback loop comprising *vri* and *pdp1*, consistent with prior models [17]. Although simplifications and assumptions were employed to reduce complexity, essential biological insights were preserved. The models use probability-based ODEs governed by mass action kinetics, as described in [2]. Additional assumptions and modeling strategies from previous work were retained, supported by relevant *in vitro* and *in vivo* evidence. It is important to emphasize that our models operate at a mesoscopic cellular scale, assuming that proteins and mRNAs are uniformly distributed and well-mixed within the system. The system volume is expressed in nanoliters' (nL) and concentrations in nanomolar (nM).

4.1 Model A

Figure 1 presents the updated conceptual framework for model A. While the core architecture remains consistent with previously described molecular networks [2, 17, 23], this revised model incorporates three interlinked feedback loops: the *cwo*, *vri/pdp1*, and *per/tim* loops. Each loop is driven by the transcriptional activator CLK/CYC. In the *per/tim* loop, CLK/CYC activates promoters containing E-box elements within the *per* and *tim* genes, leading to the production of PER and TIM proteins. These proteins form a heterodimer that binds to CLK/CYC and inhibits its own gene

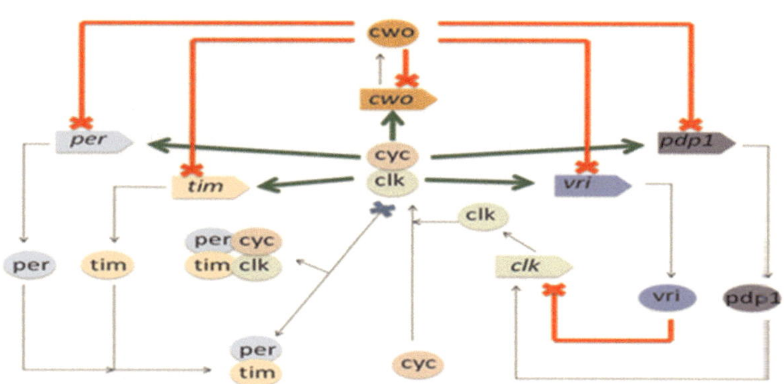

Figure 1.
Schematic diagram for model A Red lines with blunt ends represent repressive interactions, while green arrows indicate activation. Gray arrows depict the processes of transcription and translation. Genes are illustrated as rectangles with concave sides, whereas proteins and protein complexes are represented by elliptical shapes.

expression. Similarly, in the *vri/pdp1* loop, CLK/CYC attaches to the E-boxes in the *vri* and *pdp1* promoters. The resulting proteins, VRI and PDP1, interact with the V/P-box in the *clk* promoter, where they. Respectively repress and activate *clk* transcription. The third loop involves the *cwo* gene. Here, CLK/CYC initiates transcription by targeting the E-box in the *cwo* promoter. The resulting CWO protein binds to E-box elements of five core clock genes—*per*, *tim*, *vri*, *pdp1*, and *cwo*—acting as a broad repressor across the system.

To streamline Model A, several simplifying assumptions were adopted:

1. While nuclear and cytoplasmic shuttling of key clock components does occur, it was not included in the model to avoid the added complexity of explicit compartmentalization.

2. As a result of this simplification, detailed regulatory steps involving phosphorylation and dephosphorylation by kinases like DBT, CK2, and PP1 were not explicitly represented. Phosphorylation plays a pivotal role in controlling interactions and timing within the system, but the random, stochastic nature of these processes makes parameter estimation challenging. Instead, the model accounts for the net impact of phosphorylation on protein stability by incorporating degradation rates for PER and TIM. This effectively captures the influence of DBT- and PP1-mediated hyperphosphorylation, which typically targets these proteins for degradation.

3. Sequence analysis of a 4-kb region upstream of the pdp1 promoter identified six E-boxes [17]. Additional studies reported up to four functional E-boxes in the *vri* promoter and as many as five in the *per* and *tim* promoters [12, 17, 24]. For the cwo gene, up to 20 CLK/CYC-responsive E-boxes were detected in the 5′ intronic region [20]. However, due to incomplete data on transcription factor binding dynamics at these sites, the model did not include E-box counts. Instead, it assumes that CWO and CLK/CYC occupy the same E-boxes at separate times: CLK/CYC binding activates transcription, whereas CWO binding represses it. To capture this behaviour without adding unnecessary complexity, the number of E-boxes in each promoter was set to one, with probabilistic functions reflecting the stochastic dynamics.

4. In the model, CLK/CYC dimers bound to E-boxes are assumed to activate transcription without inhibition from PER/TIM heterodimers. Once CLK/CYC is engaged with the E-box, PER/TIM cannot interfere with transcription [25]. This is in contrast to mammalian systems, where CLK/BMAL1 complexes can be repressed by CRY even while remaining bound to DNA [26].

5. Experimental data from pacemaker neurons consistently show that the concentration of the CYC activator is significantly higher than that of other clock proteins [16]. Previous modelling studies accounted for this by fixing CYC's concentration at 1 nM [2]. We have retained this assumption across all three models (A, B, and C).

4.2 Model B

The rationale behind this conceptual model centers on the structural characteristics of the basic helix-loop-helix (bHLH) domain found in the CWO protein. Proteins with

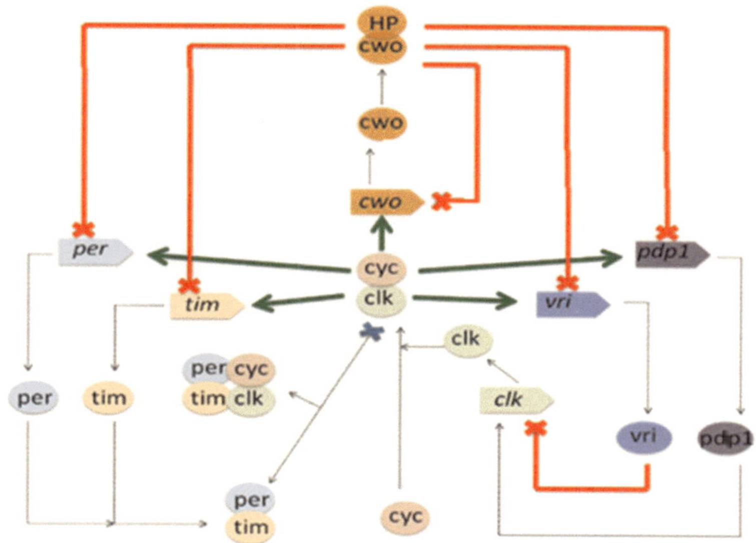

Figure 2.
*Schematic diagram for model B (conventions are given in **Figure 1**).*

this domain typically possess a 15-amino-acid basic region adjacent to the HLH motif. The HLH region primarily facilitates the formation of homo- or heterodimers, which is essential for DNA-binding repressor activity, as two basic regions are necessary to enable repression.

Proteins related to CWO that contain a hairy-orange domain often possess a WRPW motif at the C-terminus. This motif mediates interactions with corepressor proteins such as GROUCHO, enabling their function as transcriptional repressors. However, sequence analysis reveals that CWO lacks the WRPW motif, suggesting it may repress transcription independently or potentially through interactions with unidentified cofactors or domains yet to be discovered.

Drawing from experimental literature suggesting that CWO may function as a dimer [21], a revised conceptual framework—Model B—was developed (**Figure 2**). This model hypothesizes that CWO operates as a heterodimer [21]. In *Drosophila*, other bHLH-orange domain proteins have been shown to repress transcription by binding to E-boxes following heterodimer formation [27]. Based on these molecular findings, CWO is proposed to form heterodimers with candidate proteins such as Mγ, SIDE, or Mβ. Additionally, the potential for CWO homodimerization as a means of transcriptional repression has also been considered [21].

While the possibility of homodimer formation remains, supporting evidence is limited. Therefore, this model assumes the formation of a CWO heterodimer (CWD) with a hypothetical protein (HP), which could correspond to SIDE, Mβ, or Mγ as suggested in prior studies [21].

Model B retains the core structure of Model A, with the key distinction being the incorporation of CWD—formed through CWO-HP dimerization. In this revised model, CWD, rather than CWO alone, binds to E-boxes in all CLK target genes, thereby competing with CLK/CLK dimers for E-box occupancy.

4.3 Model C

All three research studies on CWO discovery [19–22] have consistently reported that CWO can function both as a transcriptional activator and repressor, although the mechanism underlying this duality remains unclear.

In *cwo*^{B9} mutant flies, particularly in large lateral ventral neurons, the null mutation does not result in arrhythmicity. However, a notable reduction in mRNA levels of E-box-regulated genes is observed. This effect is widely attributed to the loss of CWO's transcriptional activation on E-box elements [22]. While *in vivo* studies have confirmed that CWO can bind to E-boxes, its canonical role as a transcriptional repressor—consistent with its classification within the Myc-type basic helix-loop-helix (bHLH) domain family—raises questions about how it might also act as an activator. A plausible hypothesis is that CWO functions as an indirect activator by rhythmically destabilizing other transcriptional repressors that are co-expressed in phase with it.

Given that PER and TIM protein oscillations are temporally aligned with CWO expression, we investigated possible post-translational interactions among CWO, PER, and/or TIM. Since phosphorylation is a key post-translational modification, we analyzed the CWO protein sequence for DBT (Doubletime) phosphorylation sites using the GPS 2.0 (Group-based Prediction System) software [28]. This tool, despite being based on mammalian kinases, enables inference by using CK1 (the mammalian homolog of DBT) as a proxy. GPS 2.0 predicted over 25 potential CK1 binding sites within the Drosophila CWO protein.

To validate and complement these findings, we also employed NetPhosK, a neural network-based phosphorylation site prediction tool [29], which identifies potential phosphorylation at serine, threonine, and tyrosine residues. NetPhosK 2.0, trained on a broad set of experimentally validated phosphorylation sites, covers several kinases including CKII, Cam-II, PKA, PKG, PKC, Cdc2, and also predicts CK1 and GSK3 sites. The results from NetPhosK were consistent with those from GPS 2.0, confirming the presence of numerous candidate phosphorylation sites. Nevertheless, in the absence of

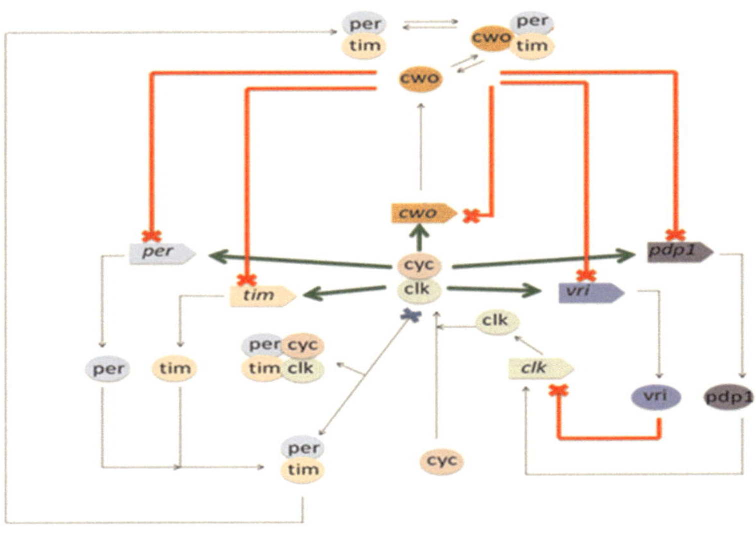

Figure 3.
The network diagram of conceptual model C.

DOI: http://dx.doi.org/10.5772/intechopen.1011916

structural data for the CWO protein, the functional relevance of these predicted sites remains uncertain.

Based on this bioinformatic evidence, we introduce a post-translational interaction component—referred to as CWPT—into model C (**Figure 3**), which involves interactions between CWO, PER, and TIM. Model C retains the core architecture of models A and B, with the added feature of phosphorylation-mediated interactions. As illustrated in **Figure 3**, CWO competes with CLK/CYC for binding to E-boxes within the promoters of *per*, *tim*, *cwo*, *pdp1*, and *vri*, similar to what is described in model A.

5. Pathway modeling

The Drosophila circadian clock includes seven known rhythmic transcription factors (TFs), with CWO being a key component. CLK/CYC heterodimers initiate the transcription of *cwo*, *pdp1*, *vri*, *tim*, and *per* by binding to E-box consensus sequences in their promoters. Rhythmic expression of the *clk* gene arises from the competitive binding of PDP1 (an activator) and VRI (a repressor) at the V/P-box of the *clk* promoter. Concurrently, PER/TIM dimers inhibit CLK/CYC activity, thereby repressing the transcription of *cwo*, *pdp1*, *vri*, *per*, and *tim*. Additionally, in models A, B, and C, CWO, CWD, and CWPT, respectively, bind to the E-boxes in the promoters of *cwo*, *vri*, *pdp1*, *per*, and *tim*, competing with CLK/CYC for binding.

There are two primary approaches for modeling transcriptional regulation. The first, widely used in earlier Drosophila circadian clock models, represents transcription rates using Hill functions or other monotonic functions. The second approach involves explicitly modeling the binding and unbinding of TFs through both forward and reverse reaction kinetics [30, 31]. This more detailed method is also commonly applied in mammalian and other circadian rhythm models [2].

Most of the foundational models of the Drosophila circadian clock simulate transcriptional regulation—including both activation and repression loops—using Hill-type or specialized mathematical functions. When parameterized appropriately, these models can replicate the molecular oscillations of core clock proteins [32, 33]. Specifically, Hill functions have been employed to model the activation of E-box-containing genes by CLK/CYC and their repression by PER/TIM, under the assumption that transcription factors bind and unbind rapidly at promoter sites. However, Hill functions are inherently limited in that they cannot account for time delays; they model protein oscillations without temporal lag.

Experimental studies indicate, for instance, that following CLK/CYC-mediated activation of *pdp1* and *vri*, there is a 3–4 hour delay before their respective mRNAs reach peak levels. Similarly, in CWO mutant backgrounds, a 2–3 hour delay is observed before corresponding proteins accumulate. While the exact mechanisms underlying these delays remain unclear, incorporating them into models is essential for biological realism—something Hill-based approaches cannot achieve.

In cases where the components responsible for such delays are not fully understood, models employing explicit equations for transcription factor binding and unbinding kinetics have successfully simulated these observed time lags [2, 30, 31]. Furthermore, Hill functions are inadequate for representing competitive interactions between transcription factors targeting the same regulatory elements—a critical limitation. In Drosophila circadian transcriptional control, such competitive dynamics are well-documented: for example, VRI and PDP1 compete for the V/P-box in the *clk* promoter [17], and CWO competes with CLK/CYC for E-box binding [21].

A prior model incorporates this competitive regulation of *clk* transcription by VRI and PDP1 using the following equation [33]:

$$R_{clk} = V_{clk}\left(\frac{[PDP1]^2}{[PDP1]^2 + K_{PDC}^2}\right)\left(\frac{K_{VC}^2}{[VRI]^2 + K_{VC}^2}\right) + R_{Cbas} \tag{1}$$

Here, R_{clk} represents the transcription rate of the *clk* gene, V_{clk} is the maximum transcription rate, and R_{Cbas} denotes the basal transcription level in the absence of transcription factor binding at the V/P-box.

In Eq. (1), Hill functions are used to model the regulatory effects of PDP1 and VRI on *clk* gene expression. The term involving K_{PDC} represents the activation of *clk* by PDP1, where K_{PDC} denotes the association constant between PDP1 and the *clk* promoter. Similarly, the repression by VRI is modeled using another Hill function, with K_{VC} representing the association constant for VRI binding to the V/P-box in the *clk* promoter. This formulation assumes a switch-like dynamic, where an increase in VRI leads to a decrease in PDP1 activity, and vice versa, effectively capturing the competition between the two transcription factors.

However, *in vitro* studies have shown that such competition does not occur in a strictly exclusive manner. Instead, both PDP1 and VRI can bind concurrently and influence *clk* transcription simultaneously [17], suggesting that the Hill-function-based representation may oversimplify the true molecular dynamics of this regulatory interaction.

It is accepted that the effective transcriptional rate of a gene—whether in its activated or repressed state—can be determined by multiplying the transcription rates associated with each state by the corresponding probabilities of the promoter being in that state [2]. This probabilistic approach to modeling transcriptional regulation offers a biologically meaningful alternative to using arbitrary switching functions. Since a promoter cannot be simultaneously active and inactive, the probability of it being in either state can be explicitly calculated.

As previously discussed, the *cwo*, *tim*, *pdp1*, *per*, and *vri* gene promoters contain multiple E-box elements. Based on this, we propose a critical hypothesis: the transcriptional repressor CWO competes with the activator CLK/CYC for binding to any available E-box within these promoters. For modelling purposes, we assume that the binding of a single E-box is sufficient to trigger either activation or repression. This forms the basis for deriving the kinetic equations in our new model, which is grounded in binding probabilities.

The competitive binding of CWO and CLK/CYC to an E-box can be described through the following reversible reactions:

$$B + A \rightleftarrows BA$$

$$B + C \rightleftarrows BC$$

Here, B denotes the E-box binding site in the gene promoter, A represents the activator complex CLK/CYC, and BA is the E-box bound by CLK/CYC. The rate constants for CLK/CYC binding and unbinding are denoted $b_{cc\_}$ and u_{bc}, respectively. Similarly, C is the repressor of CWO, and BC represents CWO bound to the E-box, with b_{cw} and ub_{cw} as the binding and unbinding rates.

Applying mass action kinetics, the dynamics of these interactions are governed by the following ordinary differential equations (ODEs):

$$\frac{d[BA]}{dt} = [B][A]b_{cc} - [BA]ub_{cc} \tag{2}$$

$$\frac{d[BC]}{dt} = [B][C]b_{cw} - [BC]ub_{cw} \tag{3}$$

If V represents the total cellular volume (in moles), then the molecular counts of the species BA, BC, and unbound B are given by $[BA]V$, $[BC]V$, and $[B]V$, respectively. Letting n denote the total number of E-box binding sites, we proceed from here to define the system based on these molecular interactions.

$$[B]V + [BA]V + [BC]V = n \tag{4}$$

$$[B]V = n - [BA]V - [BC]V \tag{5}$$

$$[B] = \left(\frac{n}{V}\right) - [BA] - [BC] \tag{6}$$

Substituting the value of $[B]$ in Eqs. (4) and (5) we get,

$$\frac{d[BA]}{dt} = \left(\left(\frac{n}{V}\right) - [BA] - [BC]\right)[A]b_{cc} - [BA]ub_{cc} \tag{7}$$

$$\frac{d[BC]}{dt} = \left(\left(\frac{n}{V}\right) - [BA] - [BC]\right)[C]b_{cw} - [BC]ub_{cw} \tag{8}$$

If Pr_{ba} and Pr_{bc} is the probability of A binding to B and C binding to B, respectively, then

$$Pr_{ba} = \frac{[BA]V}{n} \Rightarrow [BA] = \left(\frac{n}{V}\right)Pr_{ba}$$

Similarly,

$$[BC] = \left(\frac{n}{V}\right)Pr_{bc}$$

Substituting the values of $[BA]$ and $[BC]$ in Eq. (7) and (8) we obtain

$$\frac{d\left(\frac{n}{V}\right)Pr_{ba}}{dt} = \left(\left(\frac{n}{V}\right) - \left(\frac{n}{V}\right)Pr_{ba} - \left(\frac{n}{V}\right)\right)Pr_{bc}\,[A]b_{cc} - \left(\left(\frac{n}{V}\right)Pr_{ba}\right)ub_{cc} \tag{9}$$

That simplifies to,

$$\frac{dPr_{ba}}{dt} = (1 - Pr_{ba} - Pr_{bc})[A]b_{cc} - Pr_{ba}ub_{cc} \tag{10}$$

Similarly,

$$\frac{d\left(\frac{n}{V}\right)Pr_{bc}}{dt} = \left(\left(\frac{n}{V}\right) - \left(\frac{n}{V}\right)Pr_{ba} - \left(\frac{n}{V}\right)Pr_{bc}\right)[C]b_{cw} - \left(\left(\frac{n}{V}\right)Pr_{bc}\right)ub_{cw} \tag{11}$$

and this simplifies to,

$$\frac{dPr_{bc}}{dt} = (1 - Pr_{ba} - Pr_{bc})[C]b_{cw} - Pr_{bc}ub_{cw} \tag{12}$$

The E-box is assumed to be bound by CLK/CYC, and the gene transcription rate is tc_{av}. When CWO is bound as a repressor, the rate is tc_{dc} and without any molecules binding, the basal transcriptional rate is tc_{dv}.

No E-box is bound with the probability

$$(1 - Pr_{ba} - Pr_{bc})^n \tag{13}$$

whereas the probability of an E-box binding is

$$[1 - (1 - Pr_{ba} - Pr_{bc})^n] \tag{14}$$

The probability of only CLK/CYC being bound is

$$\left[\frac{(A)}{(A) + (C)}\right][1 - (1 - Pr_{ba} - Pr_{bc})^n] \tag{15}$$

Similarly, the probability of CWO being bound will be

$$\left[\frac{(C)}{(A) + (C)}\right][1 - (1 - Pr_{ba} - Pr_{bc})^n] \tag{16}$$

Consequently, the transcription rate will be

$$tc_{av}\left\{\left[\frac{(A)}{(A) + (C)}\right][1 - (1 - Pr_{ba} - Pr_{bc})^n]\right\} + tc_{dc}\left\{\left[\frac{(C)}{(A) + (C)}\right][1 - (1 - Pr_{ba} - Pr_{bc})^n]\right\}$$
$$+tc_{dv}(1 - Pr_{ba} - Pr_{bc})^n \tag{17}$$

and which simplifies to

$$\left\{\frac{tc_{av}(A) + tc_{dc}(C)}{(A) + (C)}\right\}[1 - (1 - Pr_{ba} - Pr_{bc})^n] + tc_{dv}(1 - Pr_{ba} - Pr_{bc})^n \tag{18}$$

Analogous equations can be applied to describe (1) the competitive binding between VRI and PDP1 at the *clk* promoter in models A, B, and C, and (2) the competition between CWD and CWPT with the CLK/CYC complex for E-box binding in CLK-regulated genes such as *cwo*, *vri*, *tim*, *per*, and *pdp1*. These interactions are modelled using the same set of probability-based rate equations.

In contrast to previous models that employed Michaelis-Menten kinetics [32, 33], the present models utilize mass action kinetics to describe reaction rates. This approach not only simplifies the modeling process but also reduces the number of parameters that must be estimated. The model proposed by Xie and Kulasiri [2], which is also based on mass action principles, demonstrated that its parameters are particularly sensitive to perturbations. Such sensitivity plays a critical role in model validation and in conducting comprehensive *in silico* analyses. These considerations reinforce the rationale for employing mass action kinetics in our modeling framework.

As an illustration, the following ordinary differential equation (ODE) from Model A captures the dynamic behavior of *per* mRNA and PER protein concentrations over time:

$$\frac{d(per_m)}{dt} = \left\{\left(\left[\frac{p_{43}(CC) + p_{49}(CWO)}{(CC) + (CWO)}\right][1 - (1 - Pr_{ba} - Pr_{bc})^{p_{79}}]\right) + p_{56}(1 - Pr_{ba} - Pr_{bc})^{p_{79}}\right\}$$
$$\times per_n - (p_{63} \times per_{\mathrm{m}}) \tag{19}$$

The equations expressing time evolution of proteins are governed by mass action kinetics:

$$\frac{d(PER)}{dt} = (p_{57} \times per_m) - (p_{83} \times PER \times TIM) + (p_{84} \times PT) - (p_{85} \times PER) \tag{20}$$

The rate equation presented above is incorporated into Model A to describe the temporal dynamics of PER protein. The initial term represents the translation rate of *per* mRNA, followed by terms accounting for the association and dissociation kinetics of PER/TIM (PT) dimers. The final term captures the degradation of PER protein over time.

In the system of derived ordinary differential equations (ODEs), distinct typographic conventions are employed to clearly distinguish between rate constants and variables. Protein and dimer complex names are denoted using uppercase letters, whereas mRNA species are represented using lowercase letters accompanied by the subscript ‘m’.

Specific abbreviations used in the model include: PT for the PER/TIM dimer complex, CCPT for the CLK/CYC/PER/TIM super complex, CC for the CLK/CYC dimer, CWPT for the CWO/PER/TIM interaction complex, and PDP for PDP1. All other proteins are represented using their standard three-letter biological abbreviations. Detailed descriptions of all the model parameters are provided in **Table 1** for Models A, B, and C.

The following notation is used to represent the probability of CLK/CYC (CC) transcriptional activator binding to E-box elements in various gene promoters:

- Pr_{cper}: binding in the *per* promoter
- Pr_{ct}: binding in the *tim* promoter
- Pr_{cv}: binding in the *vri* promoter
- Pr_{cpdp}: binding in the *pdp1* promoter
- Pr_{ccwo}: binding in the *cwo* promoter

Similarly, the probabilities of CWO, CWD, or CWPT binding to E-boxes are denoted as:

- Pr_{cwper}: binding in the *per* promoter
- Pr_{cwt}: binding in the *tim* promoter
- Pr_{cwv}: binding in the *vri* promoter
- Pr_{cwpdp}: binding in the *pdp1* promoter
- Pr_{cwo}: binding in the *cwo* promoter

Parameter(s)	**Kinetic rates**
p1–p5	CLK/CYC binding to per-, tim-, pdp1-, vri-, and cwo-, E-boxes, respectively
p6	PDP1 binding to V/P-box promoter in clk
p7	VRI binding to V/P-box promoter in clk
p8–p12	CWO binding to E-boxes of cwo, per, tim, pdp1, and vri, respectively
p18–p22	CLK/CYC unbinding from E-boxes of per, tim, pdp1, vri, and cwo, respectively
p23–p24	Unbinding of PDP1 and VRI from V/P-box in clk, respectively
p25–p29	CWO unbinding from E-boxes of cwo, per, tim, pdp1, and vri, respectively
p35	Association of CLK/CYC complex
p38	Association of CLK/CYC/PER/TIM complex
p39	Dissociation of CLK/CYC complex
p42	Dissociation of CLK/CYC/PER/TIM complex
p43–p48	Transcription by CLK/CYC at per, tim, pdp1, vri, clk, and cwo, respectively
p47	Transcription by PDP1 at clk, respectively
p49–p54	Transcription by CWO at per, tim, pdp1, vri, and cwo, respectively
p55	Basal transcription of clk without VRI and PDP1, respectively
p56	Basal transcription of inactive cwo, vri, per, pdp1, and tim, respectively
p57–p62	Translation of per, tim, pdp1, vri, clk, and cwo mRNAs, respectively
p63–p68	Degradation of per, tim, pdp1, vri, clk, and cwo mRNAs, respectively
p72–p74	Degradation of VRI, CLK, and CWO proteins, respectively
p76	Degradation of CLK/CYC complex, respectively
p79–p82	E-box counts in per/tim, pdp1, vri, and cwo promoters, respectively
p85–p86	Degradation of PER and TIM proteins, respectively
p94–p95 *p37*	Association/dissociation of PER/TIM dimer complex, respectively Association of CWD complex
p41	Dissociation of CWD complex
p77 *p103* *p104* *p105* *p106*	Degradation of CWD complex Degradation of PER/TIM complex Association of CWPT interaction complex Dissociation of CWPT interaction complex Degradation of CWPT interaction complex

Table 1.
Description of parameters for model A, B, and C.

Additional binding probabilities are defined as follows:

- Pr_{vc}: VRI binding to the V/P-box in the *clk* promoter
- Pr_{pc}: PDP1 binding to the V/P-box in the *clk* promoter
- Pr_{pd}: DBT binding to PER
- Pr_{pp}: PP2A binding to PER

In the ODE systems presented below, the initial conditions include: CC, CWO, CWD, PER, CCPT, CLK, TIM, PDP, VRI, CWPT, as well as transcripts clk_m, pdp_m, per_m, tim_m, vri_m, and cwo_m. CYC and HP are treated as constants.

The notation described above is consistent across all three models. For brevity, we provide only the system of ODEs corresponding to *Model A*. The equations for the other two models can be constructed similarly using parameter values listed in Table ODEs 1 and detailed in Appendices 1 and 2.

5.1 ODEs for model A

Binding Probabilities of TFs binding to regulatory elements in gene promoters

$$\frac{d(Pr_{cper})}{dt} = (1 - Pr_{cper} - Pr_{cwper}) \times p_1 \times CC - Pr_{cper} \times p_{18} \tag{21}$$

$$\frac{d(Pr_{ct})}{dt} = (1 - Pr_{ct} - Pr_{cwt}) \times p_2 \times CC - Pr_{ct} \times p_{19} \tag{22}$$

$$\frac{d(Pr_{cpdp})}{dt} = (1 - Pr_{cpdp} - Pr_{cwpdp}) \times p_3 \times CC - Pr_{cpdp} \times p_{20} \tag{23}$$

$$\frac{d(Pr_{cv})}{dt} = (1 - Pr_{cv} - Pr_{cwv}) \times p_4 \times CC - Pr_{cv} \times p_{21} \tag{24}$$

$$\frac{d(Pr_{ccwo})}{dt} = (1 - Pr_{ccwo} - Pr_{ccwo}) \times p_5 \times CC - Pr_{ccwo} \times p_{22} \tag{25}$$

$$\frac{d(Pr_{cwo})}{dt} = (1 - Pr_{ccwo} - Pr_{cwo}) \times p_8 \times CWO - Pr_{cwo} \times p_{25} \tag{26}$$

$$\frac{d(Pr_{cwper})}{dt} = (1 - Pr_{cper} - Pr_{cwper}) \times p_9 \times CWO - Pr_{cwper} \times p_{26} \tag{27}$$

$$\frac{d(Pr_{cwt})}{dt} = (1 - Pr_{ct} - Pr_{cwt}) \times p_{10} \times CWO - Pr_{cwt} \times p_{27} \tag{28}$$

$$\frac{d(Pr_{cwpdp})}{dt} = (1 - Pr_{cpdp} - Pr_{cwpdp}) \times p_{11} \times CWO - Pr_{cwpdp} \times p_{28} \tag{29}$$

$$\frac{d(Pr_{cwv})}{dt} = (1 - Pr_{cv} - Pr_{cwv}) \times p_{12} \times CWO - Pr_{cwv} \times p_{29} \tag{30}$$

$$\frac{d(Pr_{vc})}{dt} = (1 - \mathrm{Pr}_{vc} - Pr_{vc}) \times p_7 \times VRI - Pr_{vc} \times p_{24} \tag{31}$$

$$\frac{d(Pr_{pc})}{dt} = (1 - Pr_{vc} - Pr_{pc}) \times p_6 \times PDP - Pr_{pc} \times p_{23} \tag{32}$$

Time evolution of cwo, tim, vri, per, pdp1, and clk mRNA's

$$\frac{d(per_m)}{dt} = \left\{ \left(\left[\frac{p_{43}(CC) + p_{49}(CWO)}{(CC) + (CWO)} \right] [1 - (1 - Pr_{ba} - Pr_{bc})^{p_{79}}] \right) + p_{56}(1 - Pr_{ba} - Pr_{bc})^{p_{79}} \right\} \times per_n - (p_{63} \times per_m) \tag{33}$$

$$\frac{d(tim_m)}{dt} = \left\{ \left(\left[\frac{p_{44}(CC) + p_{50}(CWO)}{(CC) + (CWO)} \right] [1 - (1 - Pr_{ba} - Pr_{bc})^{p_{79}}] \right) + p_{56}(1 - Pr_{ba} - Pr_{bc})^{p_{79}} \right\}$$
$$\times tim_n - (p_{64} \times tim_m) \tag{34}$$

$$\frac{d(pdp_m)}{dt} = \left\{ \left(\left[\frac{p_{45}(CC) + p_{51}(CWO)}{(CC) + (CWO)} \right] [1 - (1 - Pr_{ba} - Pr_{bc})^{p_{80}}] \right) + p_{56}(1 - Pr_{ba} - Pr_{bc})^{p_{80}} \right\}$$
$$\times pdp_n - (p_{65} \times pdp_m) \tag{35}$$

$$\frac{d(vri_m)}{dt} = \left\{ \left(\left[\frac{p_{46}(CC) + p_{52}(CWO)}{(CC) + (CWO)} \right] [1 - (1 - Pr_{ba} - Pr_{bc})^{p_{81}}] \right) + p_{56}(1 - Pr_{ba} - Pr_{bc})^{p_{81}} \right\}$$
$$\times vri_n - (p_{66} \times vri_m) \tag{36}$$

$$\frac{d(cwo_m)}{dt} = \left\{ \left(\left[\frac{p_{48}(CC) + p_{54}(CWO)}{(CC) + (CWO)} \right] [1 - (1 - Pr_{ba} - Pr_{bc})^{p_{82}}] \right) + p_{56}(1 - Pr_{ba} - Pr_{bc})^{p_{82}} \right\}$$
$$\times cwo_m - (p_{68} \times cwo_m) \tag{37}$$

$$\frac{d(clk_m)}{dt} = \left\{ \left(\left[\frac{p_{47}(PDP) + p_{53}(VRI)}{(PDP) + (VRI)} \right] [1 - (1 - Pr_a - Pr_r)] \right) + p_{55}(1 - Pr_a - Pr_r) \right\}$$
$$\times clk_n - (p_{67} \times clk_m) \tag{38}$$

Time evolution of CWO, CLK, PER, VRI, TIM, and PDP1 proteins

$$\frac{d(PER)}{dt} = (p_{57} \times per_m) - (p_{94} \times PER \times TIM) + (p_{95} \times PT) - (p_{85} \times PER) \tag{39}$$

$$\frac{d(TIM)}{dt} = (p_{58} \times tim_m) - (p_{94} \times PER \times TIM) + (p_{95} \times PT) - (p_{86} \times TIM) \tag{40}$$

$$\frac{d(PDP)}{dt} = (p_{59} \times pdp_m) - (p_{71} \times PDP) \tag{41}$$

$$\frac{d(VRI)}{dt} = (p_{60} \times vri_m) - (p_{72} \times VRI) \tag{42}$$

$$\frac{d(CLK)}{dt} = (p_{61} \times clk_m) - (p_{35} \times CLK \times CYC) + (p_{39} \times CC) - (p_{73} \times CLK) \tag{43}$$

$$\frac{d(CWO)}{dt} = (p_{62} \times cwo_m) - (p_{74} \times CWO) \tag{44}$$

In the above equations, the first and last variables denote translational and degradation rates respectively.

The second function in Eqs. (39), (40), and (43) describes the formation of association complexes, while the third term accounts for the dissociation of these newly formed complexes.

Temporal evolution of the CLK/CYC, PER/TIM, and PER/TIM/CLK/CYC complexes

$$\frac{d(PT)}{dt} = (p_{94} \times PER \times TIM) - (p_{95} \times PT) - (p_{38} \times PT \times CC) + (p_{42} \times CCPT) - (p_{103} \times PT) \tag{45}$$

$$\frac{d(CC)}{dt} = (p_{35} \times CLK \times CYC) - (p_{39} \times CC) - (p_{38} \times PT \times CC) + (p_{42} \times CCPT) - (p_{76} \times CC) \tag{46}$$

$$\frac{d(CCPT)}{dt} = (p_{38} \times PT \times CC) - (p_{42} \times CCPT) \tag{47}$$

From the equations above, the first and second terms indicate the formation and dissociation of an association complex, respectively, while the last term represents the complex's degradation. In Eqs. (45) and (46), the third and fourth terms refer to the formation and dissociation of the respective complexes.

5.2 Parameter estimation and model implementation

To simulate our models, we utilized COPASI, an early version of VCell [34], which is a platform-independent and user-friendly biochemical simulation tool. In addition to its intuitive interface, COPASI supports the import and export of SBML codes.

In our models, all kinetic parameters required to solve the reaction ODEs were initially unknown and had to be estimated. To determine a set of parameters capable of replicating the observed biological behavior, we followed a systematic approach. In our models, none of the kinetic parameters needed to solve the reaction ODEs were known at the outset, so they all had to be estimated. To identify a set of parameters that could reproduce the observed biological behaviour, we employed a systematic procedure.

First, an initial set of parameters was estimated to roughly produce 24-hour oscillations. This was accomplished using a combination of COPASI's parameter scanning, sensitivity analysis, and parameter estimation functions. Our primary goal was to generate damped oscillations from a non-oscillatory state for all clock components. We observed that increasing the sensitivity of local parameters, particularly degradation rate parameters in the respective ODEs, facilitated the emergence of damped oscillations.

The parameter scanning function was employed by selecting a single parameter, defining its maximum and minimum values, and performing time-course simulations to observe changes in curve patterns. Two-dimensional scans were also conducted, where two parameters were independently varied across defined ranges. COPASI handled this by keeping one parameter at its minimum value while scanning the second and repeating this process across different parameter sets to enhance sensitivities. Additionally, the random distribution tool in COPASI was used to explore random parameter values that could generate oscillations. Since scanning two independent parameters with 10 intervals each required 100 time-course simulations, this was a time-intensive process.

Once damped oscillations were successfully obtained, we manually adjusted parameters to produce irregular but robust oscillations. At this stage, different parameter behaviors were analyzed, which helped identify a minimal set of parameters for estimation using COPASI's estimation algorithms. Among the various methods tested, the Levenberg–Marquardt algorithm provided the best approximate fit. However, this process demanded patience and significant computational resources.

In our investigation, the task was notably challenging, as we needed to estimate three separate models: Model A, Model B, and Model C. These models had 75, 68, and 69 dynamic parameters, respectively, in addition to their initial conditions. Fortunately, many parameters from Model A were compatible with Model B, which helped reduce the overall workload. We then validated the estimated parameters to ensure they aligned with the maximum number of biological observations. The outcomes are shown below for Models A, B, and C. It is worth mentioning that all kinetic parameters are given in units of nM per hour.

To carry out our simulations, we turned to COPASI—an early implementation of VCell [34]—a versatile and user-friendly software for simulating biochemical reactions across various platforms. COPASI's graphical interface and support for importing/exporting SBML models made it particularly suitable for our purposes.

Since none of the kinetic parameters required for the reaction ODEs in our models were initially known, they all needed to be inferred. We approached this systematically to identify parameter sets capable of reproducing the experimental biological rhythms we observed.

Our first step involved estimating an initial parameter set that could generate approximate 24-hour oscillations. We used COPASI's suite of tools—parameter scanning, sensitivity analysis, and parameter estimation—to iteratively adjust these values. Our main aim was to induce damped oscillations from non-oscillatory initial conditions for all key components of the clock network. We found that tweaking the local sensitivity of specific parameters, especially degradation rate constants, played a crucial role in achieving this.

For parameter scanning, we selected one parameter at a time and defined a range of possible values, running time-course simulations to see how this influenced the system's behavior. To explore interactions between parameters, we also performed two-dimensional scans, systematically varying two parameters within their defined ranges. COPASI handled this by fixing one parameter at its lower bound while scanning the other, repeating the process iteratively. Additionally, we used COPASI's random distribution feature to test randomly chosen parameter values that might trigger oscillatory behavior. It is worth noting that two-dimensional scans with 10 steps each required 100 time-course simulations, making this a computationally demanding process.

Once damped oscillations were established, we manually adjusted parameters further to produce irregular yet stable oscillations. At this stage, we evaluated the behaviour of each parameter, ultimately narrowing down to a minimal set of parameters suitable for estimation using COPASI's optimization algorithms. Out of several methods explored, the Levenberg–Marquardt algorithm provided the closest fit to the experimental data, though this process demanded considerable computational effort and patience.

A particular challenge was that we had to estimate parameters for three separate models—Model A, Model B, and Model C—comprising 75, 68, and 69 dynamic parameters, respectively, along with their initial conditions. Fortunately, many parameters estimated for Model A were applicable to Model B, which eased the workload. We subsequently validated the final parameter sets against the maximum number of available biological observations. The final parameter values, all expressed in nM per hour, are detailed below for Models A, B, and C.

The estimated parameters for Model A are given below as a flattened list in **Table 1**:

p1 = 0.8115; p50 = 19.1582; p2 = 0.8115; p51 = 17.5853; p3 = 0.0471; p52 = 19.2598; p4 = 0.0211; p53 = 0.1190; p5 = 0.2476; p54 = 26.8582; p6 = 12.2126; p55 = 58.8412; p7 = 0.0072; p56 = 0.0001; p8 = 0.1250; p57 = 33.8068; p9 = 0.4799; p58 = 33.8068; p10 = 0.4799; p59 = 1.9222; p11 = 0.4852; p60 = 11.5477; p12 = 0.3383; p61 = 35.7967; p18 = 0.6586; p62 = 14.2465; p19 = 0.6586; p63 = 0.0698; p20 = 0.0963; p64 = 0.0698; p21 = 0.1408; p65 = 0.0720; p22 = 0.2074; p66 = 0.0654; p23 = 2.1580; p67 = 0.6666; p24 = 0.1250; p68 = 0.0841; p25 = 22.5000; p71 = 0.1528; p26 = 10.3028; p72 = 0.9329; p27 = 10.3028; p73 = 0.0785; p28 = 10.1701; p74 = 1.0786; p29 = 14.4303; p76 = 0.0397; p35 = 2.8801; p78 = 26.6245; p38 = 90.3621; p79 = 1.0000; p39 = 0.4405; p80 = 1.0000; p42 = 2.1452; p81 = 1.0000; p43 = 18.3830; p82 = 1.0000; p44 = 18.3830; p85 = 0.4879; p45 = 18.0265; p86 = 0.4879; p46 = 61.9968; p94 = 1.1951; p47 = 119.0685; p95 = 1.4286; p48 = 1.7859; p103 = 0.0193;p49 = 19.1582.

The estimated parameters for Model B are given below:

p1 = 0.8383; p51 = 14.4091; p2 = 0.8383; p52 = 24.9594; p3 = 0.0260; p53 = 0.0019; p4 = 0.0226; p54 = 38.0013; p5 = 0.1808; p55 = 0.0011; p6 = 24.4253; p56 = 0.0001; p7 = 0.0092; p57 = 34.4307;p8 = 0.1040; p58 = 34.4307; p9 = 0.4110; p59 = 1.9325; p10 = 0.4110; p60 = 11.9584; p11 = 0.9705; p61 = 36.9362; p12 = 0.0013; p62 = 14.3208; p18 = 0.7460; p63 = 0.0683; p19 = 0.7460; p64 = 0.0683; p20 = 0.0930; p65 = 0.0787; p21 = 0.1370; p66 = 0.0620; p22 = 0.2009; p67 = 0.6809; p23 = 0.6768; p68 = 0.1330; p24 = 0.0165; p71 = 0.1505; p25 = 3.7136; p72 = 0.9827; p26 = 12.2271; p73 = 0.0613; p27 = 12.2271; p74 = 1.0794; p28 = 3.5369; p76 = 0.0496; p29 = 28.8605; p78 = 0.0001; p35 = 2.8381; p79 = 1.0000; p38 = 43.5451; p80 = 1.0000; p39 = 0.3810; p81 = 1.0000; p42 = 0.0005; p82 = 1.0000; p43 = 19.0627; p85 = 0.5740; p44 = 19.0627; p86 = 0.5740; p45 = 17.8536; p94 = 1.2806; p46 = 66.0652; p95 = 1.5048; p47 = 116.8552; p103 = 0.0319; p48 = 0.0001; p37 = 1.2385; p49 = 21.0200; p41 = 0.4795;p50 = 21.0200; p77 = 0.4798.

The estimated parameters for Model C are given below:

p1 = 0.8115; p51 = 10.6784; p2 = 0.8115; p52 = 48.3504; p3 = 0.0471; p53 = 1.3285; p4 = 0.0211; p54 = 29.8500; p5 = 0.2476; p55 = 10.4627; p6 = 12.2126; p56 = 0.0100; p7 = 0.0072; p57 = 36.0983; p8 = 0.1250; p58 = 36.0983; p9 = 0.4799; p59 = 1.5655; p10 = 0.4799; p60 = 11.1768; p11 = 0.4852; p61 = 35.0620; p12 = 0.3383; p62 = 17.4692; p18 = 0.6586; p63 = 0.0466; p19 = 0.6586; p64 = 0.0466; p20 = 0.0963; p65 = 0.0488; p21 = 0.1408; p66 = 0.1106; p22 = 0.2074; p67 = 0.6815; p23 = 2.1580; p68 = 0.1100; p24 = 0.1250; p71 = 0.1219; p25 = 22.5000; p72 = 0.8968; p26 = 10.3028; p73 = 0.0260; p27 = 10.3028; p74 = 1.3428; p28 = 10.1701; p76 = 0.0113; p29 = 14.4303; p78 = 0.4014; p35 = 2.8801; p79 = 1.0000; p38 = 47.8645; p80 = 1.0000;p39 = 0.5973; p81 = 1.0000; p42 = 0.9468; p82 = 1.0000; p43 = 16.4783; p85 = 0.2006; p44 = 16.4783; p86 = 0.2006; p45 = 18.4941; p94 = 1.3727; p46 = 81.9441; p95 = 2.8600; p47 = 122.7670; p103 = 0.1171; p48 = 2.3343; p104 = 0.0194; p49 = 6.2160; p105 = 1.1585; p50 = 6.2160; p106 = 0.0100.

The initial conditions are given below:

per_m = 0.2395; tim_m = 0.2395; pdp_m = 0.3175; vri_m = 0.2571; cwo_m = 0.2156; clk_m = 0.2583; PER = 2.7527;TIM = 2.7527; PDP = 4.1953; VRI = 3.175; CLK = 3.6628; CWO = 2.4774; PT = 0.4014; CC = 0.5566; CCPT = 0.4982; CYC = 1; CWD=1.3416; HP = 1.5632; CWPT = 1.420; Pr_{cper} = 0.0431; Pr_{ct} = 0.043; Pr_{cpdp} = 0.08; Pr_{cv} = 0.0585; Pr_{ccwo} = 0.043; Pr_{cwo} = 0.043;Pr_{cwper} = 0.0431; Pr_{cwt} = 0.043; Pr_{cwpdp} = 0.08; Pr_{cwv} = 0.0585; Pr_{vc} = 0.489; Pr_{pc} = 0.426.

We used the following values for constants:

Per_n = 0.003;Tim_n = 0.003; Clk_n = 0.003; Pdp_n = 0.003; Vri_n = 0.003; Cwo_n = 0.003.

We compared the simulation results from our models, including various test outputs, against experimental *in vitro* data. To investigate the impact of the newly introduced CWO negative feedback loop, we performed a local sensitivity analysis. The models were simulated under conditions mimicking known circadian mutants in constant darkness (DD). Additionally, we modeled light entrainment and assessed the system's robustness. All simulation outcomes under these conditions were evaluated against empirical laboratory findings. Furthermore, we proposed a hypothetical *cwoCWPT* mutant to explore how CWO might function simultaneously as an activator and repressor within pacemaker neurons.

6. Models A, B, and C: Results of simulations

Our models produced sustained oscillations of clock components under constant darkness (DD), exhibiting appropriate periods and amplitudes. Using the parameter sets listed earlier for Models A, B, and C, all three models demonstrated robust 24-hour rhythmic oscillations of mRNAs for cwo, pdp1, vri, per, clk, and tim, along with their corresponding clock proteins CWO, PDP1, VRI, PER, CLK, and TIM.

Experimental data indicate that tim, per, and cwo mRNA levels peak between circadian time (CT) 12 and CT 16 (early evening), oscillating in phase [19, 20, 22]. In contrast, clk mRNA peaks between CT 23 and CT 4 (late night to early morning) and exhibits an anti-phase relationship with tim and per mRNAs [15]. Similarly, vri and pdp1 mRNAs oscillate in anti-phase with clk mRNA but are in phase with tim and per mRNAs, peaking roughly at CT 12 to CT 14 [16–18].

Our simulated mRNA oscillations reflected these patterns: tim, per, and cwo mRNAs peaked at approximately 9.2, 9.2, and 12.7 hours in Model A; 8.9, 8.9, and 12.4 hours in Model B; and 9.2, 9.2, and 13.1 hours in Model C, respectively. Clk mRNA peaks occurred at model times of 3.8, 3.9, and 3.6 hours across the three models, consistently anti-phase with tim, per, and cwo. Vri and pdp1 mRNAs peaked at model times of 8.6 and 13 (Model A), 8.7 and 14 (Model B), and 11.2 and 11 (Model C), maintaining phase relationships observed experimentally.

Experimentally, protein peaks lag mRNA peaks by 4–6 hours for PER and TIM, with CLK protein peaking around CT 4.5, and CWO around CT 15 [20]. VRI protein peaks coincide closely with its mRNA at CT 12, while PDP1 proteins peak 3–6 hours after mRNA synthesis at approximately CT 18 [17]. The reasons for these delays remain unclear.

Our models simulated protein peaks as follows: PER, TIM, CLK, VRI, PDP1, and CWO peaked at model times of 15.3, 15.3, 4.5, 10.2, 17.8, and 14.8 hours in Model A; 15, 15, 4.5, 10.6, 18, and 13.5 hours in Model B; and 15.9, 15.9, 5, 12, 17.2, and 14.5 hours in Model C, respectively. Most simulated peaks align closely with biological circadian times, preserving the observed phase and anti-phase relationships among clock components.

Our parameter estimation aimed to generate sustained limit cycle oscillations for all circadian elements with appropriate rhythmic periods and amplitudes. All models exhibited stable oscillations near a 24-hour cycle. mRNA amplitudes ranged from 0.1 to 0.5 nM (roughly 7–34 copies), and protein concentrations ranged between 2 and 5 nM (approximately 136–340 molecules per cell). Importantly, all three models maintained these oscillations indefinitely despite the numerous unknown parameters estimated.

Slight differences in time to reach limit cycles among components caused model times to differ from biological circadian times by ± 1–3 hours in 9 out of 36 mRNA and protein oscillations. Exact circadian peak times were not essential for mutational analyses or investigating our core research questions, as relative changes in concentration, amplitude, period, and phase shifts were the focus. Nonetheless, model times were manually adjusted by ±1 to 3 hours at the start of limit cycles to better match circadian times. Peak times in circadian time for all models are in good agreement with the experimental values well within the ranges of the experimentally obtained values.

6.1 Robustness of the models to parameter variations

Drosophila circadian components can maintain phase relationships despite extrinsic noise. *In vivo* and *in vitro* studies show that environmental perturbations such as temperature and light shifts produce only minor changes in molecular oscillations. For example, shifting light exposure in flies entrained to DD by a few hours altered molecular periods by only about 0.1 hour [35]. Similarly, flies exposed to temperatures between 20 and 29°C exhibited negligible period changes (~0.06–0.2 hours). However, prolonged exposure at 20–25°C led to arrhythmicity in 20–30% of flies. Thus, an effective mathematical model must robustly maintain rhythmicity with parameter perturbations causing less than 1-hour period variation.

Global robustness analysis is impractical for systems with 65–69 transient parameters, so we applied local sensitivity analysis as in previous studies [2, 36–38]. Each parameter was individually perturbed by ±20% to assess its impact on oscillations.

For Models A, B, and C, 130, 136, and 138 local perturbation simulations were performed, respectively, excluding controls. Oscillations were maintained indefinitely across all simulations. Most parameters induced period variations under 0.4 hours, with 59, 62, and 63 parameters in Models A, B, and C, respectively, keeping period changes below 0.8 hours. This demonstrates the robustness of our models, suggesting biological resilience to environmental fluctuations.

Notably, six parameters—related to transcription and translation rates of *per* and *clk* genes and degradation rates of their mRNAs (*p43, p47, p57, p61, p63, p67*)—consistently caused period changes exceeding 0.6 hours when perturbed. Among these, a_{57}, associated with PER protein synthesis, produced the largest period deviations (>0.8 hours).

6.2 Light entrainment of the circadian clock

Following parameter estimation and model analysis described previously, we next evaluated the models' ability to predict responses to external environmental cues, or zeitgebers. While both temperature and light cycle daily, light is widely recognized as the dominant zeitgeber influencing circadian rhythms [39]. A key feature of the circadian clock is its ability to adjust to varying light conditions, particularly differing intensities and durations of exposure. This entrainment occurs through phase resetting of the circadian oscillators. In *Drosophila*, light exposure induces degradation of the TIM protein *via* its interaction with the light-activated photoreceptor CRY (cryptochrome). Genetic studies have demonstrated that the absence of TIM, such as in *tim*^{01} mutants, results in reduced cytoplasmic PER repressor levels, a phenomenon similarly observed in wild-type flies exposed to constant light [40–42]. The degradation of TIM influences the phase of oscillations of other clock proteins, leading to phase resetting. This synchronization with external cues, despite the intrinsic robustness of clock component oscillations, exemplifies the inherent stochasticity of biological systems.

The primary consequence of TIM degradation is destabilization of cytoplasmic PER, since TIM absence causes PER to remain bound to the kinase DBT, resulting in

PER hyperphosphorylation and subsequent degradation [43]. Prior modelling efforts have simulated light entrainment indirectly by increasing the degradation rate of PER protein [36, 44]. For instance, Smolen's model substituted the original PER degradation parameter with an elevated 'k_{light}' value, approximately 20% higher than baseline [33]. Other models increased TIM degradation parameters to simulate light effects [38], and Xie and Kulasiri similarly elevated TIM and PER degradation rates [2]. More recent approaches incorporated CRY mRNA cycling data to modulate TIM synthesis rates, effectively simulating light entrainment by activating CRY-dependent regulation [23]. Our models omit CRY and explicit PER/TIM phosphorylation dynamics; thus, we adopted the established approach of increasing PER and TIM degradation rates to simulate light entrainment [2].

Specifically, the parameters governing PER and TIM degradation ('$p85$' and '$p86$') were replaced by a new degradation rate termed 'k_{light}' across all three models. We simulated a 12-hour light/dark (LD) cycle, with zeitgeber time (ZT) 0 marking light onset. The light phase was defined as ZT 0 to ZT 12, and the dark phase as ZT 12 to ZT 24. During the light phase, '$p85$' and '$p86$' were replaced by an arbitrary 'k_{light}' value of 1; during the dark phase, parameters reverted to their original values, consistent with parameter estimation in constant darkness (DD). Under these LD conditions, all three models sustained rhythmic oscillations in clock proteins, maintaining expected phase and anti-phase relationships similar to DD. Comparative plots of CLK protein levels further confirmed phase resetting capability in response to light changes.

To mimic the arrhythmicity observed under constant light (LL), a higher 'k_{light}' value of 5 was used to replace the PER and TIM degradation rates. The resulting LL simulations displayed damped, arrhythmic protein oscillations consistent with experimental observations [42–45]. Collectively, these results demonstrate that all three models faithfully reproduce circadian oscillations under varying lighting conditions (DD, LD, and LL).

7. Conclusion

This study's novel contribution lies in convincingly demonstrating that CWO interacts with PER post-translationally, recapitulating biologically observed phenotypes [22]. Model C reveals that CWO functions simultaneously as both an activator and a repressor without requiring special mathematical constructs, relying solely on established biological principles. The insights from these models advance our understanding of the unique cooperative mechanisms involving CWO, representing a significant step forward for *Drosophila* circadian rhythm research. We anticipate that the validity of these models will be experimentally tested *in vivo* in the near future, to confirm which model would be better suited to explain the CWO interactions.

Appendix 1. Model B

Probabilities: TF binding to E-box or V/P-box

$$\frac{d(Pr_{cper})}{dt} = (1 - Pr_{cper} - Pr_{cwper}) \times p_1 \times CC - Pr_{cper} \times p_{18}$$

$$\frac{d(Pr_{ct})}{dt} = (1 - Pr_{ct} - Pr_{cwt}) \times p_2 \times CC - Pr_{ct} \times p_{19}$$

DOI: http://dx.doi.org/10.5772/intechopen.1011916

$$\frac{d(Pr_{cpdp})}{dt} = (1 - Pr_{cpdp} - Pr_{cwpdp}) \times p_3 \times CC - Pr_{cpdp} \times p_{20}$$

$$\frac{d(Pr_{cv})}{dt} = (1 - Pr_{cv} - Pr_{cwv}) \times p_4 \times CC - Pr_{cv} \times p_{21}$$

$$\frac{d(Pr_{ccwo})}{dt} = (1 - Pr_{ccwo} - Pr_{cwo}) \times p_5 \times CC - Pr_{ccwo} \times p_{22}$$

$$\frac{d(Pr_{cwo})}{dt} = (1 - Pr_{ccwo} - Pr_{cwo}) \times p_8 \times CWD - Pr_{cwo} \times p_{25}$$

$$\frac{d(Pr_{cwper})}{dt} = (1 - Pr_{cper} - Pr_{cwper}) \times p_9 \times CWD - Pr_{cwper} \times p_{26}$$

$$\frac{d(Pr_{cwt})}{dt} = (1 - Pr_{ct} - Pr_{cwt}) \times p_{10} \times CWD - Pr_{cwt} \times p_{27}$$

$$\frac{d(Pr_{cwpdp})}{dt} = (1 - Pr_{cpdp} - Pr_{cwpdp}) \times p_{11} \times CWD - Pr_{cwpdp} \times p_{28}$$

$$\frac{d(Pr_{cwv})}{dt} = (1 - Pr_{cv} - Pr_{cwv}) \times p_{12} \times CWD - Pr_{cwv} \times p_{29}$$

$$\frac{d(Pr_{vc})}{dt} = (1 - Pr_{vc} - Pr_{pc}) \times p_7 \times VRI - Pr_{vc} \times p_{24}$$

$$\frac{d(Pr_{pc})}{dt} = (1 - Pr_{vc} - Pr_{pc}) \times p_6 \times PDP - Pr_{pc} \times p_{23}$$

per, tim, clk, vri, cwo and pdp1 mRNA's time courses

$$\frac{d(per_m)}{dt} = \left\{\left(\left[\frac{p_{43}(CC) + p_{49}(CWD)}{(CC) + (CWD)}\right][1 - (1 - Pr_{ba} - Pr_{bc})^{p_{79}}]\right) + p_{56}(1 - Pr_{ba} - Pr_{bc})^{p_{79}}\right\} \times per_m - (p_{63} \times per_m)$$

$$\frac{d(tim_m)}{dt} = \left\{\left(\left[\frac{p_{44}(CC) + p_{50}(CWD)}{(CC) + (CWD)}\right][1 - (1 - Pr_{ba} - Pr_{bc})^{p_{79}}]\right) + p_{56}(1 - Pr_{ba} - Pr_{bc})^{p_{79}}\right\} \times tim_m - (p_{64} \times tim_m)$$

$$\frac{d(pdp_m)}{dt} = \left\{\left(\left[\frac{p_{45}(CC) + p_{51}(CWD)}{(CC) + (CWD)}\right][1 - (1 - Pr_{ba} - Pr_{bc})^{p_{80}}]\right) + p_{56}(1 - Pr_{ba} - Pr_{bc})^{p_{80}}\right\} \times pdp_n - (p_{65} \times pdp_m)$$

$$\frac{d(vri_m)}{dt} = \left\{\left(\left[\frac{p_{46}(CC) + p_{52}(CWD)}{(CC) + (CWD)}\right][1 - (1 - Pr_{ba} - Pr_{bc})^{p_{81}}]\right) + p_{56}(1 - Pr_{bc} - Pr_{bc})^{p_{81}}\right\} \times vri_n - (a_{66} \times vri_m)$$

$$\frac{d(cwo_m)}{dt} = \left\{\left(\left[\frac{p_{48}(CC) + p_{54}(CWD)}{(CC) + (CWD)}\right][1 - (1 - Pr_{ba} - Pr_{bc})^{p_{82}}]\right) + p_{56}(1 - Pr_{ba} - Pr_{bc})^{p_{82}}\right\} \times cwo_m - (p_{68} \times cwo_m)$$

$$\frac{d(clk_m)}{dt} = \left\{\left(\left[\frac{p_{47}(PDP) + p_{53}(VRI)}{(PDP) + (VRI)}\right][1 - (1 - Pr_a - Pr_r)]\right) + p_{55}(1 - Pr_a - Pr_r)\right\} \times clk_n - (p_{67} \times clk_m)$$

Time courses of PER, TIM, CLK, VRI, PDP1 and CWO proteins

$$\frac{d(PER)}{dt} = (p_{57} \times per_m) - (p_{94} \times PER \times TIM) + (p_{95} \times PT) - (p_{85} \times PER)$$

$$\frac{d(TIM)}{dt} = (p_{58} \times tim_m) - (p_{94} \times PER \times TIM) + (p_{95} \times PT) - (p_{86} \times TIM)$$

$$\frac{d(PDP)}{dt} = (p_{59} \times pdp_m) - (p_{71} \times PDP)$$

$$\frac{d(VRI)}{dt} = (p_{60} \times vri_m) - (p_{72} \times VRI)$$

$$\frac{d(CLK)}{dt} = (p_{61} \times clk_m) - (p_{35} \times CLK \times CYC) + (p_{39} \times CC) - (p_{73} \times CLK)$$

$$\frac{d(CWO)}{dt} = (p_{62} \times cwo_m) - (p_{37} \times CWO \times HP) + (p_{41} \times CWD) - (p_{74} \times CWO)$$

Time courses of PER/TIM, CWO/HP, CLK/CYC and PER/TIM/CLK/CYC complex

$$\frac{d(PT)}{dt} = (p_{94} \times PER \times TIM) - (p_{95} \times PT) - (p_{38} \times PT \times CC) + (p_{42} \times CCPT) - (p_{103} \times PT)$$

$$\frac{d(CC)}{dt} = (p_{35} \times CLK \times CYC) - (p_{39} \times CC) - (p_{38} \times PT \times CC) + (p_{42} \times CCPT) - (p_{76} \times CC)$$

$$\frac{d(CCPT)}{dt} = (p_{38} \times PT \times CC) - (p_{42} \times CCPT)$$

$$\frac{d(CWD)}{dt} = (p_{37} \times CWO \times HP) - (p_{77} \times CWD)$$

Appendix 2. Model C

Probabilities: TFs binding to E-box or V/P-box

$$\frac{d(Pr_{cper})}{dt} = (1 - Pr_{cper} - Pr_{cwper}) \times p_1 \times CC - Pr_{cper} \times p_{18}$$

$$\frac{d(Pr_{ct})}{dt} = (1 - Pr_{ct} - Pr_{cwt}) \times p_2 \times CC - Pr_{ct} \times p_{19}$$

$$\frac{d(Pr_{cpdp})}{dt} = (1 - Pr_{cpdp} - Pr_{cwpdp}) \times p_3 \times CC - Pr_{cpdp} \times p_{20}$$

$$\frac{d(Pr_{cv})}{dt} = (1 - Pr_{cv} - Pr_{cwv}) \times p_4 \times CC - Pr_{cv} \times p_{21}$$

$$\frac{d(Pr_{ccwo})}{dt} = (1 - Pr_{ccwo} - Pr_{cwo}) \times p_5 \times CC - Pr_{ccwo} \times p_{22}$$

$$\frac{d(Pr_{cwo})}{dt} = (1 - Pr_{ccwo} - Pr_{cwo}) \times p_8 \times CWO - Pr_{cwo} \times p_{25}$$

$$\frac{d(Pr_{cwper})}{dt} = (1 - Pr_{cper} - Pr_{cwper}) \times p_9 \times CWO - Pr_{cwper} \times p_{26}$$

$$\frac{d(Pr_{cwt})}{dt} = (1 - Pr_{ct} - Pr_{cwt}) \times p_{10} \times CWO - Pr_{cwt} \times p_{27}$$

$$\frac{d(Pr_{cwpdp})}{dt} = (1 - Pr_{cpdp} - Pr_{cwpdp}) \times p_{11} \times CWO - Pr_{cwpdp} \times p_{28}$$

$$\frac{d(Pr_{cwv})}{dt} = (1 - Pr_{cv} - Pr_{cwv}) \times p_{12} \times CWO - Pr_{cwv} \times p_{29}$$

$$\frac{d(Pr_{vc})}{dt} = (1 - Pr_{vc} - Pr_{pc}) \times p_{7} \times VRI - Pr_{vc} \times p_{24}$$

$$\frac{d(Pr_{pc})}{dt} = (1 - Pr_{vc} - Pr_{pc}) \times p_{6} \times PDP - Pr_{pc} \times p_{23}$$

Time evolution of per, tim, clk, vri, cwo and pdp1 mRNAs

$$\frac{d(per_m)}{dt} = \left\{\left(\left[\frac{p_{43}(CC) + p_{49}(CWO)}{(CC) + (CWO)}\right][1 - (1 - Pr_{ba} - Pr_{bc})^{p_{79}}]\right) + p_{56}(1 - Pr_{ba} - Pr_{bc})^{p_{79}}\right\}$$
$$\times per_m - (p_{63} \times per_m)$$

$$\frac{d(tim_m)}{dt} = \left\{\left(\left[\frac{p_{44}(CC) + p_{50}(CWO)}{(CC) + (CWO)}\right][1 - (1 - Pr_{ba} - Pr_{bc})^{p_{79}}]\right) + p_{56}(1 - Pr_{ba} - Pr_{bc})^{p_{79}}\right\}$$
$$\times tim_m - (a_{64} \times tim_m)$$

$$\frac{d(pdp_m)}{dt} = \left\{\left(\left[\frac{p_{45}(CC) + p_{51}(CWO)}{(CC) + (CWO)}\right][1 - (1 - Pr_{ba} - Pr_{bc})^{p_{80}}]\right) + a_{56}(1 - Pr_{ba} - Pr_{bc})^{p_{80}}\right\}$$
$$\times pdp_m - (p_{65} \times pdp_m)$$

$$\frac{d(vri_m)}{dt} = \left\{\left(\left[\frac{p_{46}(CC) + p_{52}(CWO)}{(CC) + (CWO)}\right][1 - (1 - Pr_{ba} - Pr_{bc})^{p_{81}}]\right) + p_{56}(1 - Pr_{ba} - Pr_{bc})^{p_{81}}\right\}$$
$$\times vri_m - (p_{66} \times vri_m)$$

$$\frac{d(cwo_m)}{dt} = \left\{\left(\left[\frac{p_{48}(CC) + p_{54}(CWO)}{(CC) + (CWO)}\right][1 - (1 - Pr_{ba} - Pr_{bc})^{p_{82}}]\right) + p_{56}(1 - Pr_{ba} - Pr_{bc})^{p_{82}}\right\}$$
$$\times cwo_m - (p_{68} \times cwo_m)$$

$$\frac{d(clk_m)}{dt} = \left\{\left(\left[\frac{p_{47}(PDP) + p_{53}(VRI)}{(PDP) + (VRI)}\right][1 - (1 - Pr_{a} - Pr_{r})]\right) + p_{55}(1 - Pr_{a} - Pr_{r})\right\}$$
$$\times clk_m - (p_{67} \times clk_m)$$

Time evolution of PER, TIM, CLK, VRI, PDP1 and CWO proteins

$$\frac{d(PER)}{dt} = (p_{57} \times per_m) - (p_{94} \times PER \times TIM) + (p_{95} \times PT) - (p_{104} \times PER \times TIM \times CWO)$$
$$+ (p_{105} \times CWPT) - (p_{85} \times PER)$$

$$\frac{d(TIM)}{dt} = (p_{58} \times tim_m) - (p_{94} \times PER \times TIM) + (p_{95} \times PT) - (p_{104} \times PER \times TIM \times CWO)$$
$$+ (p_{105} \times CWPT) - (p_{86} \times TIM)$$

$$\frac{d(PDP)}{dt} = (p_{59} \times pdp_m) - (p_{71} \times PDP)$$

$$\frac{d(VRI)}{dt} = (p_{60} \times vri_m) - (p_{72} \times VRI)$$

$$\frac{d(CLK)}{dt} = (p_{61} \times clk_m) - (p_{35} \times CLK \times CYC) + (p_{39} \times CC) - (p_{73} \times CLK)$$

$$\frac{d(CWO)}{dt} = (p_{62} \times cwo_m) - (p_{104} \times PER \times TIM \times CWO) + (p_{105} \times CWPT) - (p_{74} \times CWO)$$

Time evolution of PER/TIM, CLK/CYC and PER/TIM/CLK/CYC complex

$$\frac{d(PT)}{dt} = (p_{94} \times PER \times TIM) - (p_{95} \times PT) - (p_{38} \times PT \times CC) + (p_{42} \times CCPT) - (p_{103} \times PT)$$

$$\frac{d(CC)}{dt} = (p_{35} \times CLK \times CYC) - (p_{39} \times CC) - (p_{38} \times PT \times CC) + (p_{42} \times CCPT) - (p_{76} \times CC)$$

$$\frac{d(CCPT)}{dt} = (p_{38} \times PT \times CC) - (p_{42} \times CCPT) - (p_{78} \times CCPT)$$

$$\frac{d(CWPT)}{dt} = (p_{104} \times CWO \times PER \times TIM) - (p_{105} \times CWPT) - (p_{106} \times CWPT)$$

Author details

Don Kulasiri* and Jeevabharathi Ranganathan
Centre for Advanced Computational Solutions (C-fACS), Lincoln University, Christchurch, New Zealand

*Address all correspondence to: don.kulasiri@lincoln.ac.nz

References

[1] Helfrich Forster C. Neurobiology of the fruit fly's circadian clock. Genes, Brain and Behavior. 2005;**4**(2):65-76

[2] Xie Z, Kulasiri D. Modelling of circadian rhythms in drosophila incorporating the interlocked PER/TIM and VRI/PDP1 feedback loops. Journal of Theoretical Biology. 2007;**245**(2): 290-304

[3] Konopka RJ, Benzer S. Clock mutants of Drosophila melanogaster. Proceedings of the National Academy of Sciences. 1971;**68**(9):2112-2116

[4] Bargiello TA, et al. Restoration of Circadian Behavioural Rhythms by Gene Transfer in Drosophila. Nature. 1984; **312**:752-754

[5] Thomas JB et al. Molecular genetics of the single-minded locus: A gene involved in the development of the drosophila nervous system. Cell. 1988;**52**(1):133-141

[6] Hoffman EC et al. Cloning of a factor required for activity of the ah (dioxin) receptor. Science. 1991;**252**(5008): 954-958

[7] Huang ZJ et al. PER protein interactions and temperature compensation of a circadian clock in drosophila. Science. 1995;**267**(5201): 1169-1172

[8] Myers MP et al. Positional cloning and sequence analysis of the drosophila clock gene, timeless. Science. 1995; **270**(5237):805-808

[9] Gekakis N et al. Isolation of timeless by PER protein interaction: Defective interaction between timeless protein and long-period mutant PERL. Science. 1995; **270**(5237):811-815

[10] Sehgal A et al. Loss of circadian behavioral rhythms and per RNA oscillations in the drosophila mutant timeless. Science. 1994;**263**(5153): 1603-1606

[11] Hardin PE et al. Feedback of the drosophila period gene product on circadian cycling of its messenger RNA levels. Nature. 1990;**343**(6258):536-540

[12] McDonald MJ, Rosbash M. Microarray analysis and organization of circadian gene expression in drosophila. Cell. 2001;**107**(5):567-578

[13] Allada R et al. A mutant drosophila homolog of mammalian clock disrupts circadian rhythms and transcription of period and timeless. Cell. 1998;**93**(5): 791-804

[14] Darlington TK et al. Closing the circadian loop: CLOCK-induced transcription of its own inhibitors per and tim. Science. 1998;**280**(5369): 1599-1603

[15] Bae K et al. Circadian regulation of a drosophila homolog of the mammalian clock gene: PER and TIM function as positive regulators. Molecular and Cellular Biology. 1998;**18**(10):6142-6151

[16] Glossop NR et al. Interlocked feedback loops within the drosophila circadian oscillator. Science. 1999; **286**(5440):766-768

[17] Cyran SA et al. Vrille, Pdp1, and dClock form a second feedback loop in the drosophila circadian clock. Cell. 2003;**112**(3):329-341

[18] Glossop NR et al. VRILLE feeds back to control circadian transcription of clock in the drosophila circadian oscillator. Neuron. 2003;**37**(2):249-261

[19] Kadener S et al. Clockwork Orange is a transcriptional repressor and a new

drosophila circadian pacemaker component. Genes & Development. 2007;**21**(13):1675-1686

[20] Lim C et al. Clockwork orange encodes a transcriptional repressor important for circadian-clock amplitude in drosophila. Current Biology. 2007; **17**(12):1082-1089

[21] Matsumoto A et al. A functional genomics strategy reveals clockwork orange as a transcriptional regulator in the drosophila circadian clock. Genes and Development. 2007;**21**(13): 1687-1700

[22] Richier B et al. The clockwork orange drosophila protein functions as both an activator and a repressor of clock gene expression. Journal of Biological Rhythms. 2008;**23**(2):103-116

[23] Fathallah-Shaykh HM et al. Mathematical model of the drosophila circadian clock: Loop regulation and transcriptional integration. Biophysical Journal. 2009;**97**(9):2399-2408

[24] Blau J, Young MW. Cycling vrille expression is required for a functional drosophila clock. Cell. 1999;**99**(6): 661-671

[25] Yue H et al. Insights into the behaviour of systems biology models from dynamic sensitivity and identifiability analysis: A case study of an NF-Î°B signalling pathway. Molecular BioSystems. 2006;**2**(12):640-649

[26] Lee C et al. Posttranslational mechanisms regulate the mammalian circadian clock. Cell. 2001;**107**(7): 855-867

[27] Davis RL, Turner DL. Vertebrate hairy and enhancer of split related proteins: Transcriptional repressors regulating cellular differentiation and embryonic patterning. Oncogene. 2001; **20**(58):8342-8357

[28] Xue Y et al. GPS 2.0, a tool to predict kinase-specific phosphorylation sites in hierarchy. Molecular & Cellular Proteomics. 2008;**7**(9):1598-1608

[29] Blom N et al. Prediction of post-translational glycosylation and phosphorylation of proteins from the amino acid sequence. Proteomics. 2004; **4**(6):1633-1649

[30] Vilar JM et al. Mechanisms of noise-resistance in genetic oscillators. Proceedings of the National Academy of Sciences. 2002;**99**(9):5988-5992

[31] Forger DB, Peskin CS. A detailed predictive model of the mammalian circadian clock. Science Signaling. 2003; **100**(25):14806

[32] Ueda HR et al. Robust oscillations within the interlocked feedback model of drosophila circadian rhythm. Journal of Theoretical Biology. 2001;**210**(4): 401-406

[33] Smolen P et al. Modeling circadian oscillations with interlocking positive and negative feedback loops. The Journal of Neuroscience. 2001;**21**(17):6644-6656

[34] Mendes P. GEPASI: A software package for modelling the dynamics, steady states and control of biochemical and other systems. Computer Applications in the Biosciences: CABIOS. 1993;**9**(5):563-571

[35] Levine JD et al. Resetting the circadian clock by social experience in Drosophila melanogaster. Science. 2002; **298**(5600):2010-2012

[36] Lema MA et al. Delay model of the circadian pacemaker. Journal of Theoretical Biology. 2000;**204**(4): 565-573

[37] Leloup J-C, Goldbeter A. Modeling the molecular regulatory mechanism of circadian rhythms in drosophila. BioEssays. 2000;**22**(1):84

[38] Leloup J-C, Goldbeter A. Toward a detailed computational model for the mammalian circadian clock. Proceedings of the National Academy of Sciences. 2003;**100**(12):7051-7056

[39] Aschoff JR et al. Re-entrainment of circadian rhythms after phase-shifts of the Zeitgeber. Chronobiologia. 1974; **2**(1):23-78

[40] Zerr D et al. Circadian fluctuations of period protein immunoreactivity in the CNS and the visual system of drosophila. The Journal of Neuroscience. 1990;**10**(8):2749-2762

[41] Vosshall LB et al. Block in nuclear localization of period protein by a second clock mutation, timeless. Science-AAAS-Weekly Paper Edition-including Guide to Scientific Information. 1994; **263**(5153):1606-1608

[42] Price J et al. Suppression of PERIOD protein abundance and circadian cycling by the drosophila clock mutation timeless. The EMBO Journal. 1995; **14**(16):4044

[43] Shafer OT et al. Flies by night: Effects of changing day length on Drosophila's circadian clock. Current Biology. 2004;**14**(5):424-432

[44] Olde Scheper T et al. A mathematical model for the intracellular circadian rhythm generator. The Journal of Neuroscience. 1999;**19**(1):40-47

[45] Qiu J, Hardin PE. Per mRNA cycling is locked to lights-off under photoperiodic conditions that support circadian feedback loop function. Molecular and Cellular Biology. 1996; **16**(8):4182-4188

Chapter 2

An Accurate and Robust Numerical Solver for Second-Order Ordinary Boundary Value Problems Based on Continuous Genetic Algorithms: Theory, Application, and Convergence Analysis

Zaer Salem Abo-Hammour

Abstract

Continuous genetic algorithms (CGAs), previously developed by the author, are introduced in this chapter as accurate and robust numerical solvers for second-order ordinary boundary value problems. The solution methodology is based on representing each derivative in the ordinary boundary value problem by its finite difference approximation. After that, the overall residue for all unknown nodes in the given problem is determined where the solution to the boundary value problem is finally converted into an optimization problem of minimizing the overall residue or maximizing the fitness function. To demonstrate the efficiency of the algorithm and confirm its performance, eight second-order ordinary boundary value problems are included in this work covering the linearity (linear/nonlinear), singularity (singular/nonsingular), and number of equations (single equation/system of equations) scenarios. In addition to that, a convergence analysis is performed which include a genetic-related analysis and problem-related analysis. Numerical results show that CGA is an efficient and robust method for solving ordinary boundary value problems. The obtained accuracy for the solutions using CGA outperforms the results obtained using other well-known methods such as residual power series method, reproducing kernel Hilbert space method, and the homotopy analysis method.

Keywords: continuous genetic algorithms, computational intelligence, evolutionary computation, differential equations, ordinary boundary value problems; numerical solutions, singular and nonsingular problems, linear and nonlinear problems

1. Introduction

Ordinary boundary value problems (BVPs) occur frequently in engineering [1–4], heat and fluid flow [5–8], physiology [9, 10], biology [11], physics [12, 13], control and optimization theory [14], and economics [15]. Ordinary boundary value problems are classified in terms of linearity (linear/nonlinear), singularity (singular/nonsingular), and number of equations (single equation /system of equations). The solution methods of BVPs are classified into two categories which include the analytical methods yielding the exact solutions for the problem at hand subject to round off error, and the numerical methods yielding approximate solutions for the problem. Analytical methods can solve limited categories of boundary value problems which are generally linear, nonsingular, and of limited number of equations. However, if the encountered problem, as in the case of most real-life cases, is highly nonlinear, singular, and consists of a system of equations, then numerical methods are preferred in this scenario due to the limitations of the analytical methods. Based on that, researchers continuously develop new methods or modify existing methods to obtain robust, accurate, and fast numerical methods [16–22].

Numerical methods have three key performance measures: accuracy, robustness, and finally convergence speed. These criteria are critical for selecting the appropriate method in specific applications. In general, there are trade-offs between these measures where some numerical methods might excel in one measure while underperforming in others. For the solution of second-order ordinary BVPs, methods such as the finite difference method (FDM), finite element method (FEM), and shooting method are commonly employed [23–26]:

1. Accuracy: These methods are generally accurate, especially when the problem is well-behaved (linear and simple nonlinear BVP). However, accuracy may deteriorate or degrade when dealing with complex nonlinear systems of equations or singular systems unless adjustments are made. In these cases, the standard methods are often modified with techniques like mesh refinement for FEM or higher-order discretization schemes for FDM. When nonlinearity increases, perturbation techniques or the use of asymptotic methods can help maintain accuracy.

2. Robustness. It refers to the method's ability to handle a wide spectrum of problems especially highly coupled systems of equations or singular systems. Methods that are accurate for simple cases may fail when applied to singular or ill-conditioned problems. For example, FDM though powerful for well-conditioned and linear problems may need to be adapted or combined with other techniques such as Richardson extrapolation to improve robustness in more complex systems. For problems with singularities, shooting methods may struggle without significant modification by employing homotopy continuation or multiple shooting methods. For FEM, a combination of adaptive mesh refinement and higher-order basis functions is required to enhance its robustness while dealing with highly complex systems. Furthermore, both FDM and shooting methods may fail if the problem is very complex.

3. Convergence speed. Some methods, like FEM may converge quickly in simple, linear problems but may require modifications (e.g., adaptive step-sizing or preconditioned iterative methods) to when applied to highly nonlinear boundary value problems which add a computational burden and slow down the method.

Based on the previous discussion, existing numerical methods may require modifications within the standard numerical methods, or be combined with other numerical methods, if possible, to handle the three key performance measures while solving the most sophisticated case representing highly coupled nonlinear singular systems of ordinary boundary value problems. This means that the researcher should have a wealth of information and knowledge about methods' refinements and adjustments from one side and a rigid background about other numerical methods in order to deal with all categories of ordinary boundary value problems. Therefore, researchers need an insight into newer methods that might offer better performance measures than the existing numerical methods.

Genetic algorithms (GAs) are global numerical optimization methods based on the concepts of genetics and natural selection. GAs imitate nature with their Darwinian survival-of-the-fittest approach [27, 28]. This approach allows GAs to explore new directions in the search space leading to improved performance by efficiently pruning the search space. Their basic principle is the maintenance of an evolving population of solutions to the problem at hand based on the triangle of genetic operations: selection, crossover, and mutation. Throughout the genetic operations, the fitness of the individuals in the population pool is gradually improved until the convergence is achieved according to certain convergence criteria. Genetic Algorithms (GAs) are different from calculus-based methods which often rely on gradients, Hessians, or other derivative-based calculations. Instead, GAs use principles inspired by natural evolution, such as selection, crossover, and mutation, to iteratively search for optimal solutions in a problem space. This simplicity in mechanism makes GAs powerful in certain contexts [27, 28].

Continuous genetic algorithms are variant of GAs where smooth or continuous operators are used to avoid sharp jumps in the values of the optimized variables [29]. They are recommended when the variables to be optimized are coupled or correlated with each other. CGAs ideally fit for problems that have solution curves in one-dimensional space or solution surfaces in two-dimensional space [29]. Following their novel development in 2002, CGAs have been successfully applied in various fields of science and engineering. They have been used for solving linear and nonlinear ordinary boundary value problems [30–33] and singular ordinary boundary value problems [34, 35]. CGAs have been extended for the solution of linear and nonlinear partial differential equations [36–38]. Furthermore, they have been successfully applied in cartesian path generation of robot manipulators [39–42] and the solution of optimal control problems [43, 44].

In this chapter, CGAs are introduced as novel solvers for all of the combinations of classifications for the ordinary boundary value problems including linearity (linear/nonlinear), singularity (singular/nonsingular), and number of equations (single equation /system of equations). The application of CGAs for the solution of the second-order ordinary boundary value problems is triggered by the following distinct advantages over traditional numerical methods:

1. Robustness. CGAs excel in robustness because they handle both linear and nonlinear, singular and nonsingular boundary value problems without needing algorithmic modifications. As a result, they are of versatile nature. On the other hand, traditional numerical methods often require customization or augmentation when transitioning between different problem types.

2. Simplicity. This approach is simple to understand and implement since it neither resorts to advanced mathematical tools nor requires a rigid background about other numerical methods in order to deal with all categories of ordinary boundary value problems. On the other hand, traditional numerical methods often demand a strong mathematical background, especially in cases of complex, nonlinear, or singular BVPs. CGAs rely on population-based optimization principles, which can be implemented using relatively straightforward operations like selection, crossover, and mutation, making them accessible to a wider audience of engineers and researchers.

3. Global convergence. CGAs exhibit global convergence properties, meaning they explore the entire solution space and have a higher chance of avoiding local minima or divergence. On the other hand, traditional numerical methods may suffer from local convergence or divergence particularly for nonlinear, singular, or ill-conditioned problems.

4. Parallel Computation. The computational burden of CGAs can be significantly reduced by implementing them on parallel computing architectures, making them suitable for real-time applications utilizing the intrinsic population-based nature of CGA.

The organization of this chapter is as follows: in Section 2, the second-order BVPs is formulated as an optimization problem based on the fitness function. Section 3 describes CGAs in detail. Numerical results for the given ordinary boundary value problems are covered in Section 4. Comparison with other well-known numerical methods for the solution of the given ordinary boundary value problems will be discussed in Section 5, while convergence analysis is provided in Section 6. Finally, concluding remarks are presented in Section 7.

2. General formulation of the ordinary boundary value problems

In this section, the most general formulation of the ordinary boundary value problems will be considered which represent a system of ordinary differential equations of boundary type. Given the system of m two-point second-order BVPs described the following set of ordinary differential equations:

$$\begin{aligned}
&y_1''(x) = f_1(x, y_1(x), y_2(x), \ldots, y_m(x), y_1'(x), y_2'(x), \ldots, y_m'(x)),\\
&y_2''(x) = f_2(x, y_1(x), y_2(x), \ldots, y_m(x), y_1'(x), y_2'(x), \ldots, y_m'(x)),\\
&\vdots\\
&y_m''(x) = f_m(x, y_1(x), y_2(x), \ldots, y_m(x), y_1'(x), y_2'(x), \ldots, y_m'(x)),
\end{aligned} \tag{1}$$

subject to the boundary conditions

$$\begin{aligned} & y_1(a) = \alpha_1, y_1(b) = \beta_1, \\ & y_2(a) = \alpha_2, y_2(b) = \beta_2, \\ & \vdots \\ & y_m(a) = \alpha_m, y_m(b) = \beta_m, \end{aligned} \tag{2}$$

where $a \le x \le b$, α_k, β_k are real finite constants and f_k are nonlinear functions of y_k and y'_k, $k = 1, 2, \dots, m$.

The formulation process of the ordinary boundary value problems consists of the following steps:

2.1 Interval discretization

The solution interval $[a, b]$ is discretized in this step into a number of equally spaced nodes or mesh points by setting

$$\begin{aligned} x_i &= a + ih \\ i &= 0, 1, \dots, N \\ h &= \frac{b - a}{N} \end{aligned} \tag{3}$$

Thus, at the interior mesh points, x_i, $i = 1, 2, \dots, N - 1$, is actually following equation Eq. (3) and does not require a separate equation number

$$y''_k(x_i) = f_k(x_i, y(x_i), y'(x_i)) = 0, \quad x_1 \le x_i \le x_{N-1}, \tag{4}$$

subject to the boundary conditions

$$y(x_0) = \alpha, y(x_N) = \beta,$$

where $y = (y_1, y_2, \dots, y_m)$, $\alpha = (\alpha_1, \alpha_2, \dots, \alpha_m)$, and $\beta = (\beta_1, \beta_2, \dots, \beta_m)$, $k = 1, 2, \dots, m$.

2.2 Finite difference approximations of the derivatives

Following the discretization step, all derivatives appearing in the system of ordinary boundary value problems are replaced by their equivalent finite difference approximations. Different variations of finite difference approximations are reported in literature including the forward finite difference formulas, the backward finite difference formulas, and the hybrid forward finite difference formulas which include a mix of forward and backward formulas. The central finite difference formulas are special cases of the hybrid formulas [45].

The *(R + 1)-point* forward difference formula of the *kth* derivative is represented according to the following equation:

$$y^k_F(x_i) = g(x_i, x_{i+1}, \dots, x_{i+k}, \dots, x_{i+R}) \tag{5}$$

The $(L + 1)$-*point* backward difference formula of the *kth* derivative is represented according to the following equation:

$$y_B^k(x_i) = g(x_{i-L}, \dots, x_{i-K}, \dots, x_{i-1}, x_i) \tag{6}$$

The *(L + R + 1)-point* hybrid difference formula of the *kth* derivative is represented according to the following equation:

$$y_H^k(x_i) = g(x_{i-L}, \dots, x_i, \dots, x_{i+R}) \tag{7}$$

A *special case* of the hybrid difference formula is the *(2 L + 1)-point* central difference formula when $R = L$ (number of nodes to the left of the node of interest equals number of nodes to the right of the node of interest) which is represented according to the following equation:

$$y_C^k(x_i) = g(x_{i-L}, \dots, x_i, \dots, x_{i+L}) \tag{8}$$

These approximation formulas are applied for $y'(x_i)$ and $y''(x_i)$, $i = 1, 2, \dots, N-1$, for all interior mesh points. The accuracy of the formulas can be improved by increasing the number of points included in the formulas which can go up to N points.

2.3 Conversion into a system of algebraic equations

After the derivative replacement, the system of differential equations is transformed into a system of linear/nonlinear algebraic equations in the following form:

$$F_k\left(y(x_{i-L}), y\left(x_{i-(L+1)}\right), \dots, y(x_{i-1}), y(x_i), y(x_{i+1}), \dots, y(x_{i+R})\right) = g_k(x_i) k = 1, 2, \dots, m \tag{9}$$

2.4 Residual calculations

The residual of any general interior unknown node, $i = 1, 2, \dots, N-1$, is given as

$$\mathrm{r}_k(i) = F_k\left(y(x_{i-L}), y\left(x_{i-(L+1)}\right), \dots, y(x_{i-1}), y(x_i), y(x_{i+1}), \dots, y(x_{i+R})\right) - g_k(x_i) = 0 \tag{10}$$

The overall individual residue, R, is a function of the residuals of all interior nodes is defined as

$$\mathrm{R} = \sqrt{\sum_{k=1}^{m} \sum_{i=1}^{N-1} (\mathrm{r}_k(i))^2} \tag{11}$$

2.5 Fitness function calculations

In GAs, fitness functions are used to guide the selection of better solutions over generations. Since GAs are designed to maximize the fitness function, it is required to transform the minimization problem of minimizing the overall individual residual, R, into maximization of the fitness function, F, as given in the following equation:

$$F = \frac{\delta}{\delta + R} \tag{12}$$

where δ is a small positive number that is taken as a unity in this work.

The individual fitness is inversely proportional to the overall individual residual with optimal values of $R = 0$ and $F = 1$.

3. Description of the continuous genetic algorithm

There are several key considerations that should be taken into account while designing and implementing GAs for optimization. These considerations greatly influence the effectiveness of GAs:

1. *Representation of problem solutions*. This factor involves defining how the problem's potential solutions or individuals are represented. Commonly, solutions are encoded as binary strings or real-valued vectors depending on the problem's nature.

2. *Generation of the initial population*. The initial population can be generated randomly or seeded with known solutions or heuristics. This impacts how quickly the GA converges to optimal solutions. *Random initialization* ensures a wide diversity of solutions but may require more generations to converge, while *heuristic initialization* seeds the population with better initial solutions, which can speed up convergence but risks limiting exploration if too focused. A balance between exploration (diversity) and exploitation (using good initial solutions) is essential.

3. Design of genetic operators. Genetic operators like crossover and mutation define how solutions are modified and how new candidate solutions are generated. These operators must align with the chosen genetic representation and problem domain.

4. Formulation of the Fitness Function. Fitness function is crucial for evaluating and guiding the evolution of solutions toward the optimal result. It must be carefully designed to suit the specific optimization problem. If the fitness function is poorly selected, then GA might not be able to reach to optimal or near optimal solutions, and even might fail.

5. Genetic parameters. Parameters like population size, mutation rate, crossover probability, and termination criteria are essential for controlling the performance and behavior of the GA. Poor parameter tuning can lead to suboptimal performance or convergence issues.

The previous considerations have indeed led to the development of numerous *variants of GAs*. These variants are tailored to suit different types of optimization problems and performance requirements, making GAs highly versatile [29, 30]. Based on that, the nature of the optimization problem being addressed directly influences the choice of the GA variant:

1. *Correlation between unknown variables*

   - If the *unknown variables* in the optimization problem are *uncorrelated or independent from each other*, then conventional (discrete) GA often works well, as it allows independent evolution of each *variable*.

   - If the *unknown variables are correlated or dependent on each other*, where a change in one variable impacts others, then CGA is better suited, as it can capture these interdependencies more effectively.

2. *Smoothness of the solution curve*

   - If the resulting solution curves requires *smoothness*, CGAs are preferred, as they maintain gradual and smooth changes across the solution space.

   - In cases where *non-smooth* or abrupt changes are acceptable, then conventional (discrete) GA can efficiently handle such abrupt changes.

CGA is often chosen when smoothness in the solution is necessary or when the *unknown variables* are tightly coupled or *correlated with each other*. CGA uses real-valued encodings and operators suited for continuous optimization problems. The reader is kindly asked to refer to [29, 30, 39, 40] in order to know more details about CGA. The CGA used in this work, as given in **Figure 1**, consists of the following steps.

1. *Initialization:* In the context of CGAs, ensuring that the initialization function is smooth and satisfies the given boundary conditions is essential. Two initialization functions are used in this research including the modified normal Gaussian function (MNGF) and the modified tangent hyperbolic function (MTHF). MNGF is given in the following equation:

$$p_j(k,i) = r(k,i) + A \times \exp\left(-0.5\left(\frac{i-\mu}{\sigma}\right)^2\right)\sin\left(\frac{\pi}{N}i\right) \tag{13}$$

While MTHF is given in the following equation:

$$p_j(k,i) = r(k,i) + A \times \tanh\left(\frac{i-\mu}{\sigma}\right)\sin\left(\frac{\pi}{N}i\right) \tag{14}$$

for each $i = 1, 2, \ldots, N-1, j = 1, 2, \ldots, N_p$, and $k = 1, 2, \ldots, m$.

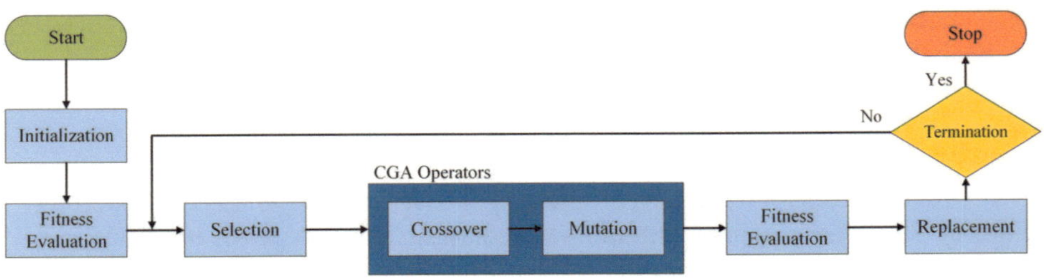

Figure 1.
Flowchart of continuous genetic algorithm.

Where

$p_j(k,i)$ is the *ith* nodal value of the *kth* variable $(y_k(x_i))$ for the *jth* parent,

$r(k,i)$ is the ramp function of the *ith* nodal value of the *kth* variable and defined as

$$r(k,i) = \alpha_k + \frac{\beta_k - \alpha_k}{N} i \tag{15}$$

μ *and* σ are random numbers within the range $[1, N-1]$ and $\left[0, \frac{N-1}{3}\right]$, respectively.

N_p is the population size.

$\sin\left(\frac{\pi}{N} i\right)$ is the corrector function that might result in an overshoot/undershoot in the initialization function which might exceed the values of the given boundary conditions at some interior mesh points but not at the boundary point $\{a, b\}$.

It is to be noted that MTHF is convex in a subinterval of the independent variable and concave in the remaining interval, while MNGF is either convex or concave within the entire the solution interval $[a, b]$.

The controlling parameters of the two initialization functions are the random numbers A, μ, and σ. Randomness is based on the fact that the initial population should be diverse since the required solutions are not known. The mentioned diversity is the key parameter in having an information-rich initial population.

It is to be noted that the controlling parameters describe the following:

i. A specifies the amplitude of the corrector function $\left\{ \sin\left(\frac{\pi}{N} i\right) \right\}$

ii. σ specifies degree of dispersion of the corrector function $\left\{ \sin\left(\frac{\pi}{N} i\right) \right\}$.

iii. μ specifies the center of the MNGF, while μ specifies the intersection point between the ramp function and the MTHF, which determines the convexity point.

The range of A depends on the boundary conditions α_k and β_k, $k = 1, 2, \ldots, m$ according to the following guidelines:

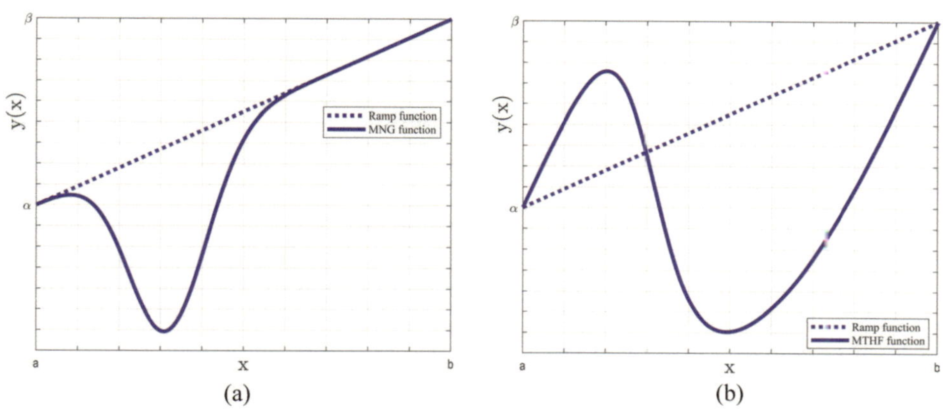

Figure 2.
(a) Modified normal Gaussian function (MNGF), (b) modified tangent hyperbolic function (MTHF).

i. within the range $\{[-3|\beta_k - \alpha_k|, 3|\beta_k - \alpha_k|] \}$ if $\beta_k - \alpha_k \neq 0$.

ii. within the range $[-3\alpha_k, 3\alpha_k]$ if $\beta_k = \alpha_k$.

iii. within the range $\left[-\frac{N-1}{3}, \frac{N-1}{3}\right]$ if $\beta_k = 0$ and $\alpha_k = 0$.

The two initialization functions are given in **Figure 2**.

2. *Fitness evaluation:* In GAs, *fitness evaluation* is a critical step that determines how well each individual in the population performs with respect to the problem being solved. The fitness function quantifies the "goodness" of a solution, guiding the selection of individuals for reproduction as given in Eq. (12).

3. *Selection:* Selection process is a fundamental mechanism in GAs that determines which individuals will contribute their genetic material or traits to the next generation. The goal of selection is to ensure that individuals with higher fitness —those that represent better solutions—are more likely to pass on their characteristics (genes) to future generations. By favoring "fitter" individuals, the overall quality of the population improves over time, steering the search toward optimal solutions.

4. *Crossover:* Crossover plays a pivotal role in sharing valuable information between individuals, facilitating the discovery of better solutions by recombining the parents' genetic material to create new individuals (offspring) that may exhibit traits from both parents, potentially leading to better solutions. The choice of crossover method and its configuration are critical to the success of the algorithm, influencing both the diversity of the population and the speed of convergence toward an optimal solution. In CGA, crossover is given according to the following equation:

$$c_l(k,i) = c(k,i)p_s(k,i) + (1 - c(k,i))p_h(k,i), c_{l+1}(k,i) = (1 - c(k,i))p_s(k,i) \\ +c(k,i)p_h(k,i), c(k,i) = 0.5\left(1 + \tanh\left(\frac{i-\mu}{\sigma}\right)\right), \tag{16}$$

for each $i = 1, 2, \ldots, N-1$ and $k = 1, 2, \ldots, m$,
where

p_s and p_h represent the two selected parents from the selection process, c_l and c_{l+1} are the two children generated from by the crossover process, c represents the crossover smooth weighting function within the range $[0, 1]$. Parameters μ and σ are as given in the initialization process.

It is worth mentioning that the crossover weighting function is the tangent hyperbolic function that has initial values that are close to zero and final value that are close to unity. This means that there will be a gradual transition between the values of the first and the second parents. According to **Figure 3**, the generated children are described as follows:

i. The initial part of the first child is similar to the initial part of the second parent, while final part of the first child is similar to the final part of the first parent.

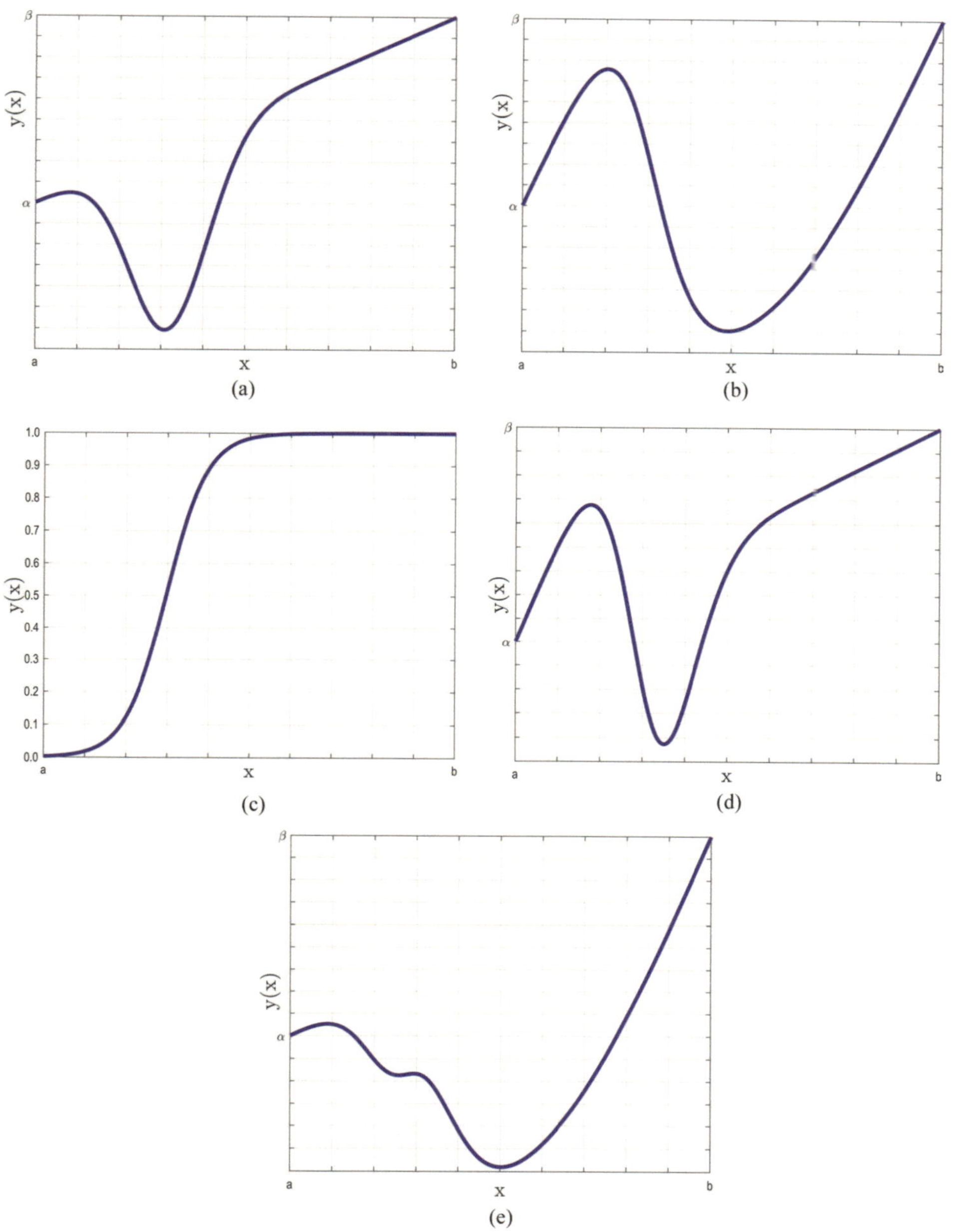

Figure 3.
Crossover process of single equation BVP: (a) First Parent, (b) second parent, (c) crossover function, (d) first child, and (e) second child.

ii. The initial part of the second child is similar to the initial part of the first parent, while final part of the second child is similar to the final part of the second child.

iii. The intermediate part of both children is a nonlinear mixture between the values of the first and second parents according to the tangent hyperbolic function.

As compared with conventional genetic algorithm, crossover is performed in a sharp manner (using single point crossover, multipoint crossover, or uniform crossover schemes) in conventional genetic algorithm without any transition period, while it is performed in a gradual smooth manner in CGA. However, information exchange between the two parents is still performed. The crossover process is shown in **Figure 3**.

5. *Mutation:* Mutation is a vital genetic operator in GAs that introduces random alterations to individual solutions (or chromosomes) in the population. It serves as a mechanism to maintain diversity by *introducing variability* to individuals and prevent the algorithm from prematurely converging to a suboptimal solution, which can occur when the gene pool becomes homogeneous due to selection and crossover processes. This means that mutation is crucial for maintaining a balance between exploration (searching new areas) and exploitation (refining known good solutions). The mutation process, given in **Figure 4**, is governed by the following formulas:

$$m_j(k,i) = c_j(k,i) + Am(k,i), m(k,i) = \exp\left(-0.5\left(\frac{i-\mu}{\sigma}\right)^2\right)\sin\left(\frac{\pi}{N}i\right), \qquad (17)$$

for each $i = 1, 2, \ldots, N-1, j = 1, 2, \ldots, N_p$, and $k = 1, 2, \ldots, m$,
where

$m_j(k,i)$ is the ith variable value of the kth solution curve for the mutated jth child for.

$c_j(k,i)$ is the ith variable value of the kth solution curve for the jth child produced through the crossover process,

$m(k,i)$ is the modified Gaussian mutation function,

A is as given in the initialization process.

Regarding the mutation center $\{\mu\}$, three methods are used for generating the mutation center where each of the three method is applied to one-third of the population to maintain diversity: -

i. *Deterministic* method. This means that the mesh point with the maximum absolute residual is chosen as the mutation center.

ii. *Lamarckian* method. The mutation center is found in a probabilistic manner where mesh points with larger values of the residual error have more chances of being selected as the mutation center.

iii. *Random* method. In this method, the mutation center is chosen randomly.

Regarding the dispersion factor $\{\sigma\}$, two methods are used for generating the dispersion factor where each of the two method is applied to one-half of the population to maintain diversity: -

i. *Knowledgeable* method. This method is applied according to the following steps: -

a. Select two random integer numbers, μ_L and μ_R such that $\mu_L \le \mu \le \mu_R$ and both of μ_L and μ_R lie within the range $(1, N-1)$. These two numbers

DOI: http://dx.doi.org/10.5772/intechopen.1008305

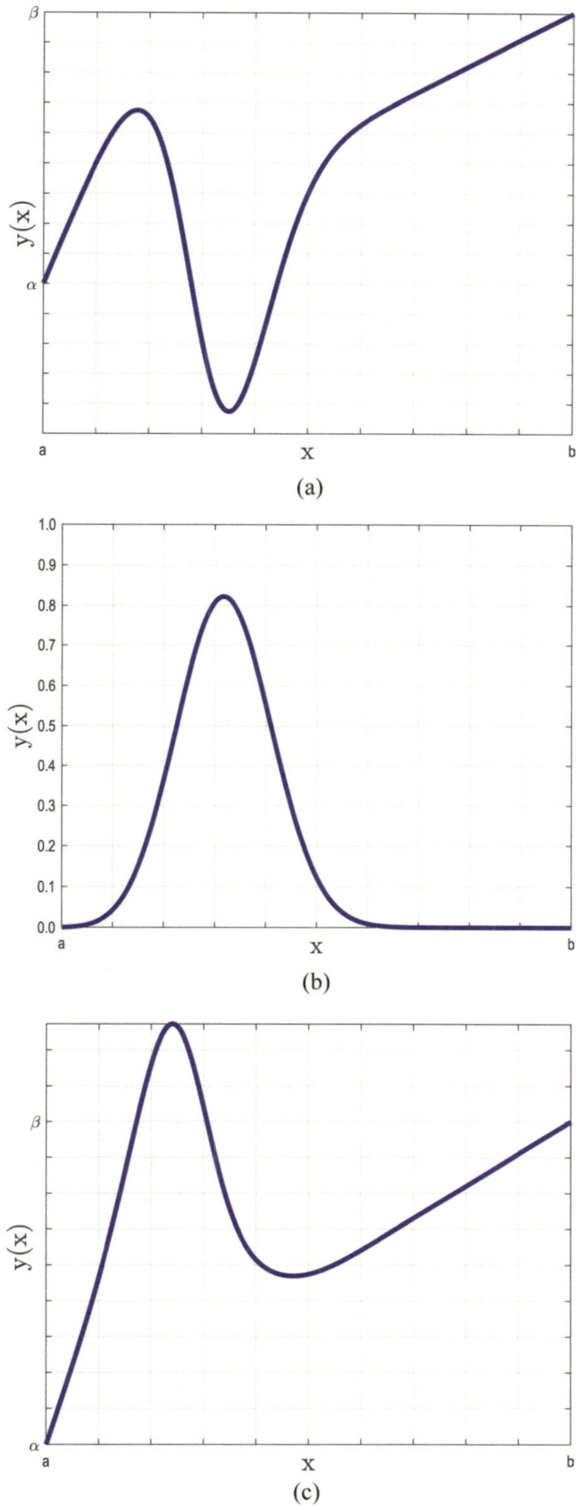

Figure 4.
Mutation process. (a) Second child. (b) Mutation function. (c) Mutated child.

represent the two random numbers to the left and right of the mutation center respectively that is previously found according to the methods of generating the mutation center $\{\mu\}$.

b. Calculate the average error function, $E_f(i)$, at all nodes within the range $\mu_L \leq i \leq \mu_R$.

c. Calculate the ratio between the calculated average error, $E_f(i)$, and the error at the mutation center, $E_f(\mu)$.

d. Finally, the dispersion factor, σ, is chosen as a random number that is a function of both the calculated ratio and the value $\{\mu_R - \mu_L\}$.

ii. *Random* method. In this method, the dispersion factor is chosen randomly.

6. *Replacement.* Replacement marks the end of one generation and the beginning of the next, closing the "life cycle" of the population and preparing it for the next round of genetic operations. After generating offspring through the application of genetic operators (such as selection, crossover, and mutation), the next task is to determine how the new individuals (offspring) will interact with the existing population (parents). The replacement process is crucial for maintaining the evolutionary dynamics of the algorithm and influences its ability to explore and exploit the solution space effectively.

7. *Termination. Termination* determines when the algorithm should stop running. Establishing appropriate termination criteria is essential to ensure that the algorithm concludes after finding a satisfactory solution, avoiding unnecessary additional computation, and preventing premature termination. The termination criteria that are used in CGA include the following:

1. *Fitness threshold.* The algorithm terminates when the fitness of the best individual in the population meets or exceeds a predefined threshold value, or when $\{F_{best} \geq F_{thr1}\}$. This criterion ensures that the GA stops once a solution of sufficient quality is found. This approach is useful when the optimal fitness value is known beforehand which is unity in the current fitness formulation, that is, $\{F_{optimal} = 1\}$.

2. *Residual threshold.* In this criterion, the algorithm terminates when the maximum nodal residual of the best individual in the population is below a specified threshold. This criterion ensures that the solution satisfies the necessary conditions for convergence. This approach is useful when the optimal residual value is known beforehand which is zero in the current fitness formulation, that is, $\{\sum_{k=1}^{m} \mathrm{r}_k(i) \leq r_{thr}\}$.

3. *Maximum generations.* Another straightforward termination criterion is to stop the algorithm after reaching a maximum number of generations

$\{gen \geq gen_{max}\}$. This approach limits the computational effort and ensures that the algorithm does not run indefinitely. In this case, the algorithm terminates regardless of the quality of the solution found, that is, the algorithm might converge to optimal solution or not.

4. *Improvement in fitness.* The algorithm can also terminate when the improvement in the fitness value of the best individual of the population over a specified number of generations falls below a specified threshold $\{F_{best}(gen) - F_{best}(gen - gen_{thr}) < F_{thr2}\}$. This criterion indicates that the population has stagnated and further generations are unlikely to yield significant improvements. The algorithm terminates regardless of the quality of the solution found, that is, the algorithm might converge to optimal solution or not.

After terminating the algorithm, the solution to the second-order ordinary boundary value is the best individual so far found where the unknown nodal values for all curves are the solutions obtained. If the termination conditions are not met, then the algorithm will go back to step 2 or the fitness evaluation step.

To enhance the performance of the Continuous Genetic Algorithm, two additional operators are introduced. These operators aim to improve the quality of solutions and maintain diversity within the population, addressing issues such as stagnation and premature convergence.

1. Elitism. Elitism is a technique used to preserve the best-of-generation individual(s) of the size $\{N_{elite}\}$ from the current generation and ensure their survival into the next generation without alteration. This operator achieves the following goals: -
   i. Preservation of best solution(s). The best individual or several top-performing individuals are copied directly from the parent population into the next generation without alteration. This ensures that valuable information is not lost during the process of selection, crossover, and mutation, as it can sometimes happen when only offspring are retained.

   ii. *Monotonic improvement*: Elitism guarantees that the fitness of the best solution in the population will not decrease across generations. As a result, the fitness function behaves as a *monotonically non-decreasing* function, meaning that the algorithm never loses progress by discarding highly fit individuals.

   iii. *Balance between exploration and exploitation*. By ensuring that the best individuals survive to the next generation, elitism promotes exploitation of known good solutions while allowing other individuals in the population to explore new possibilities.

2. Extinction and Immigration. Extinction and Immigration is introduced in CGA to deal with situations of stagnation or when the population becomes homogeneous, thereby reducing diversity and halting progress. This operator works by performing a significant reset of the population to reintroduce diversity and reinvigorate the search process by replacing all or part of the population with new, randomly generated individuals. This operator consists of two stages:

i. Extinction stage. During the extinction stage, all individuals are removed from the current population except for the best-of-generation individual (s) of the size $\{N_{ext}\}$. This ensures that the best solution(s) found so far is preserved, while the rest of the population is cleared to make room for new genetic material.

ii. Mass-immigration stage. After the extinction stage, the mass-immigration stage introduces new genetic material by filling the population with new, randomly generated individuals of the size $\{N_p - N_{ext}\}$ individuals to keep the population size fixed. The generated population in this stage is divided into two equal parts each of $N_p/2$ size;

   a. First part, with $\{j = N_{ext} + 1, \dots, N_p/2\}$, is produced as in the initialization phase.

   b. Second part with $\left\{j = \frac{N_p}{2} + 1, \dots, N_p\right\}$ is produced by executing continuous mutation to the best-of-generation individual(s) as given by the formula

$$p_j(k,i) = p_1(k,i) + Am(k,i)\,, \tag{18}$$

for each $i = 1, 2, \dots, N-1, j = \frac{N_p}{2} + 1, \frac{N_p}{2} + 2, \dots, N_p$, and $k = 1, 2, \dots, m$, where

$p_j(k,i)$ represents the i-th variable value of the k-th curve for the j-th parent generated using immigration operator,

p_1 is any of the best-of-generation individual(s) $\{1, \dots, N_{ext}\}$,

m represents the Gaussian mutation function,

A is a random number as given in the initialization process.

The evolution process of the continuous genetic algorithm for solving an ordinary boundary value problem can be summarized as follows. An individual is a candidate solution for the ordinary boundary value problem that consists of m curves (representing a system of m differential equations where $m = 1$ for single differential equation) each of $\{m(N-1)\}$ unknown nodal values. This means that the total number of evolving unknown nodes within CGA is equal to $\{m(N-1)\}$. The population of individuals undergoes the selection process which results in a mating pool. Pairs of individuals which are chosen from the mating pool are crossed over with probability p_{ci}. Within the selected pair of parents, individual solution curves are then crossed with probability p_{cc}. This crossover produces new offspring, which represent combinations of the parent solutions. After that, every child generated through cross-over undergoes mutation with probability p_{mi}. Furthermore, each *individual solution curve* within the child is mutated with probability p_{mc}, introducing random changes to the nodal values to maintain diversity in the population. After mutation, the next generation is produced using a *replacement strategy*, which may involve replacing part or all of the parent population with the offspring population. The entire process—selection, crossover, mutation, and replacement—is repeated across multiple generations until the algorithm terminates when a predefined convergence criterion is met.

The *m curves* of the best individual at the point of convergence represent the solution to the boundary value problem. This means that the final goal of finding the required nodal values is translated into finding the fittest individual in genetic terms.

Problem number	Linearity condition	Number of equations	Singularity condition	Reference used
Problem 1	Linear	Single	Nonsingular	[30]
Problem 2	Linear	Single	S	[34]
Problem 3	Linear	System	Nonsingular	[31]
Problem 4	Linear	System	S	[35]
Problem 5	Nonlinear	Single	Nonsingular	[30]
Problem 6	Nonlinear	Single	S	[34]
Problem 7	Nonlinear	System	Nonsingular	[31]
Problem 8	Nonlinear	System	S	[35]

Table 1.
Eight selected ordinary boundary value problems.

4. Numerical results using CGA

In this section, all possible combinations of second-order ordinary boundary value problems are included covering the linearity (linear/nonlinear), singularity (singular/nonsingular), and number of equations (single equation /system of equations) scenarios. This results in a total number of eight different problems. The classification information about the eight selected ordinary boundary value problems is given in **Table 1**, while their full details and description are given in **Table 2**. The solution of this set of boundary value problems will verify the computational efficiency of CGA including the robustness of the method from one side and its accuracy from the other side. The last column of **Table 1** shows the reference from which the given problem is selected.

Throughout the algorithm, the genetic operators are applied according to the following procedures:

a. Initialization method: 50% of the initial population is created using the MNGF, while the other 50% is created by MTHF.

b. Selection scheme. Rank-based selection is the default selection method.

c. Replacement strategy. Generational replacement is applied.

d. Termination criteria. CGA is stopped when one of the following conditions is met:

Fitness threshold. $F_{best} \geq 0.999999$.
Residual threshold. $\sum_{k=1}^{m} r_k(i) \leq 10^{-8}$.
Maximum generations. $gen \geq 3000$.
Improvement in fitness. $F_{best}(gen) - F_{best}(gen - 500) < 10^{-3}$.

The solution of every problem is the average of 12 runs to remove any bias in the solutions due to probabilistic nature of CGA. Additional CGA-related numerical parameters are given in **Table 3**.

The eight problems were solved using CGA without any failure to solve any of them as shown in **Table 4**. The average number of generations required for convergence varies from about 800 generations up to about 1600 generations. The average fitness of the final solution achieves at least 0.99999 value (five nines). The average absolute

Problem Number	Boundary value problem	Boundary conditions	Exact solution(S)
1	$y'' = 2y' - y - 3$	$0 \le x \le 1$ $y(0) = -3$ $y(1) = -2.264241$	$y(x) = 2xe^{(x-2)} - 3$
2	$y'' + \frac{1}{x}y' + y = x^2 - x^3 - 9x + 4$	$0 \le x \le 1$ $y(0) = 0$ $y(1) = 0$	$y(x) = x^2 - x^3$
3	$y_1''(x) = xy_2'(x) + \cos(\pi x)y_1(x) - y_2(x) + f_1(x)$ $f_1(x) = \cosh(x) - \frac{1}{2}\sin(2\pi x) - x\sinh(x) - \pi^2\sin(\pi x)$ $y_2''(x) = -y_1'(x) + y_2'(x) - \sinh(x)y_1(x) + f_2(x)$ $f_2(x) = \cosh(x) + \sinh(x)(\sin(\pi x) - 1) + \pi\cos(\pi x)$	$0 \le x \le 1$ $y_1(0) = 0$ $y_1(1) = 0$ $y_2(0) = 1$ $y_2(1) = \cosh(1)$	$y_1(x) = \sin(\pi x)$ $y_2(x) = \cosh(x)$
4	$y_1'' + \frac{2x-1}{x^2\sqrt{x}}\{y_1' + x^2y_2'\} - \frac{x^3}{\cos(\frac{\pi}{2}x)}\{y_1 - \exp(x)y_2\} + f_1(x) = 0$ $y_2'' + \frac{x}{(x-1)^2}y_1' - \frac{x}{\sin(\pi x)}\{y_1 - xy_2\} + f_2(x) = 0$ $f_1(x)$ and $f_2(x)$ are not included here due to their length and can be found simply by using the exact solutions.	$0 \le x \le 1$ $y_1(0) = 0$ $y_1(1) = 0$ $y_2(0) = 1$ $y_2(1) = 0$	$y_1(x) = \sin(\pi x)$ $y_2(x) = x^2 - x$
5	$y'' = \frac{1}{8}(32 + 2x^3 - yy')$	$1 \le x \le 2$ $y(1) = 17$ $y(2) = 12$	$y(x) = x^2 + \frac{16}{x}$
6	$y'' + \frac{60}{\sqrt{x}(x-1)^2}y' + \frac{3}{\tan(x)}\cos(y) = f(x$ $\boldsymbol{f(x)} = \frac{3\cos(\sin(\pi x) + \exp(1))}{\tan(x)} - \pi^2 \sin(\pi x) + \frac{60\pi\cos(\pi x)}{\sqrt{x}(x-1)^2}$	$0 \le x \le 1$ $y(0) = exp(1)$ $y(1) = exp(1)$	$y(x) = sin(\pi x) + exp(1)$
7	$y_1''(x) = \frac{2}{1+2y_2(x)}\left((y_1'(x))^2 + (1-2x)^2\right) + f_1(x)$ $f_1(x) = -\frac{16x^2-16x+4}{2\exp\left(-\frac{x}{\sqrt{2}}\right)+1} - 2$ $y_2''(x) = -\cosh(y_2'(x)) + \frac{1}{2}y_1(x) + \ln(y_2(x)) + f_2(x)$ $f_2(x) = \cosh\left(\frac{1}{\sqrt{2}}\exp\left(-\frac{x}{\sqrt{2}}\right)\right) + \frac{1}{2}\exp\left(-\frac{x}{\sqrt{2}}\right) + \frac{x^2}{2} + \left(\frac{1}{\sqrt{2}} - \frac{1}{2}\right)x - \frac{1}{4}$	$0 \le x \le 1$ $y_1(0) = \frac{1}{2}$ $y_1(1) = \frac{1}{2}$ $y_2(0) = 1$ $y_2(1) = \exp\left(-\frac{1}{\sqrt{2}}\right)$	$y_1(x) = x - x^2 + \frac{1}{2}$ $y_2(x) = \exp\left(-\frac{x}{\sqrt{2}}\right)$

DOI: http://dx.doi.org/10.5772/intechopen.1008305

Problem Number	Boundary value problem	Boundary conditions	Exact solution(S)
8	$y_1'' + \frac{20}{x(x-1)}\left\{y_1' + (y_2')^2\right\} - \frac{\cos(x)}{\sinh^2(x)}\left\{y_1^2 + x\sin(y_1 y_2)\right\} + f_1(x) = 0$ $y_2'' + \frac{5\exp(x)}{x\sin(x)}(y_2')^3 - \frac{x}{\sqrt{1-x}}\left\{\sinh(x)(y_2)^3 + \sin(x) y_2 \cos(y_1)\right\} + f_2(x) = 0$ $f_1(x)$ and $f_2(x)$ are not included here due to their length and can be found simply by using the exact solutions.	$0 \le x \le 1$ $y_1(0) = 1$ $y_1(1) = exp(1)$ $y_2(0) = 0$ $y_2(1) = \sinh(1)$	$y_1(x) = \exp(x)$ $y_2(x) = \sinh(x)$

Table 2.
Full details and description about the eight selected ordinary boundary value problems.

Parameter	Description
$N_p = 500$	Population size
$p_{ci} = 0.9$	Individual crossover probability
$p_{mi} = 0.9$	Individual mutation probability
$p_{cc} = 0.5$	Curve crossover probability
$p_{mc} = 0.5$	Curve mutation probability
$R_{br} = 0.1$	Rank-based ratio
$N_{elite} = 10\% \times N_p$	Percentage of elite parents that are passed to the next generation

Table 3.
CGA-related parameters.

residual values for all unknown nodes within each problem lies within the range of 10^{-6} and 10^{-11}. Finally, and most importantly, the average absolute error for all unknown nodes within each problem lies within the range of 10^{-8} and 10^{-12}. Based on this information, it is clear that CGA is an accurate method for the second-order ordinary boundary value problems from one side and a robust method from the other side.

Numerical solutions for the eight ordinary boundary value problems are given in **Tables 5–16**. The tables show the nodal value (x_i), the exact value of the solution according to **Table 2** (y_{exact}), the solution value using CGA (y_{CGA}), the average absolute residual value for all unknown nodes in the given problem excluding the boundaries $\overline{(|r(i)|)}$, and the average absolute error for all unknown nodes in the given problem excluding the boundaries $\overline{(|e(i)|)}$.

5. Comparison between CGA and other well-known methods

Following the successful implementation of the CGA for the solution of the given boundary value problems, some of the given problems are solved using other well-known numerical methods for comparison purposes. The modern numerical methods that are used for comparison with CGA include the homotopy analysis (HA) method [16], reproducing kernel Hilbert space (RKHS) method [17], and the residual power series (RPS) method [19]. These methods are suited for linear systems. However, when solving the nonlinear systems, these methods require some major modifications.

Problems 3 and 7 representing a system of two equations are selected for numerical comparisons. These problems are solved using the three methods. **Tables 17** and **18** show a comparison between the average absolute error $\overline{(|e_1(i)|)}$ of CGA together with other aforementioned methods for problem 3, while **Tables 19** and **20** show the comparison for problem 7.

Table 21 gives the average absolute error $\overline{(|e(i)|)}$ for all unknown nodes of all variables of the third and seven problems using HA, RKHS, RPS, and CGA methods.

The following facts are deduced from **Table 21**:

1. CGA is the best method for the solution of the two problems where $\overline{(|\boldsymbol{e}(\boldsymbol{i})|)}$ is the lowest.

2. For the linear and nonlinear cases, $\overline{(|e(i)|)}$ is relatively of the same order between 10^{-8} and 10^{-9} using CGA.

Problem number	Average number of generations	Average fitness	Average absolute residual	Average absolute error
1	783	0.999999836	$4.93755096 \times 10^{-9}$	$8.40372698 \times 10^{-10}$
2	889	0.999999000	$5.20529873 \times 10^{-11}$	$3.73551730 \times 10^{-12}$
3	1084	0.999998600	$1.62682994 \times 10^{-6}$	$1.15150865 \times 10^{-8}$
4	1597	0.999999836	$3.30061098 \times 10^{-7}$	$2.16394062 \times 10^{-8}$
5	947	0.999999148	$2.76917868 \times 10^{-9}$	$4.71315066 \times 10^{-10}$
6	1227	0.999990970	$3.36319589 \times 10^{-7}$	$2.43196999 \times 10^{-9}$
7	1256	0.999999000	$4.02015967 \times 10^{-7}$	$5.65247209 \times 10^{-9}$
8	1649	0.999999741	$2.09437978 \times 10^{-8}$	$9.38055433 \times 10^{-10}$

Table 4.
Convergence data of the eight boundary-value problems.

x_i	y_{exact}	y_{CGA}	$\overline{\lvert r(i)\rvert}$	$\overline{\lvert e(i)\rvert}$
0.0	−3.0000000000000	−3.0000000001736	0	0
0.1	−2.9700862761555	−2.9700862754116	$4.37036222 \times 10^{-9}$	$7.43837000 \times 10^{-10}$
0.2	−2.9338804447114	−2.9338804452087	$2.92201996 \times 10^{-9}$	$4.97328700 \times 10^{-10}$
0.3	−2.8903898855684	−2.8903898865460	$5.74402589 \times 10^{-9}$	$9.77634980 \times 10^{-10}$
0.4	−2.8384827856043	−2.8384827847279	$5.14934436 \times 10^{-9}$	$8.76420000 \times 10^{-10}$
0.5	−2.7768698398516	−2.7768698384557	$8.20124247 \times 10^{-9}$	$1.39585400 \times 10^{-10}$
0.6	−2.7040836432701	−2.7040836422433	$6.03285627 \times 10^{-9}$	$1.02679400 \times 10^{-10}$
0.7	−2.6184554897524	−2.6184554904323	$3.99456385 \times 10^{-9}$	$6.79876000 \times 10^{-10}$
0.8	−2.5180892609405	−2.5180892600729	$5.09719346 \times 10^{-9}$	$8.67543900 \times 10^{-10}$
0.9	−2.4008320493435	−2.4008320488454	$2.92635016 \times 10^{-9}$	$4.98065700 \times 10^{-10}$
1.0	−2.2642411176571	−2.2642411183519	0	0

Table 5.
Numerical results for problem 1.

3. For the nonlinear problem, the HA methods failed, while the accuracy of the RKHS method and the RPS method fall much below that of CGA.

Based on the above discussion, it can be easily observed that CGA outperforms the other methods in terms of accuracy and robustness.

However, as CGA is a population-based approach, its computational burden when applied on sequential computers is relatively large. For example, the average number of generations required for convergence for the eight problems given in **Table 4** is equal to 1179 generation. Keeping in mind that the population size is $N_p = 500$ individuals as given in **Table 3**, then the total number of candidate solutions that are evaluated using the CGA is of the order of 10^5. This computational burden of CGA can be significantly reduced by implementing them on parallel computing architectures, making them suitable for real-time applications utilizing the population-based nature

x_i	y_{exact}	y_{CGA}	$\overline{\lvert r(i)\rvert}$	$\overline{\lvert e(i)\rvert}$
0.0	0	0	0	0
0.1	0.009	0.0089999999973	$4.86658042 \times 10^{-11}$	$2.68724626 \times 10^{-12}$
0.2	0.032	0.0319999999954	$4.67374089 \times 10^{-11}$	$4.58882932 \times 10^{-12}$
0.3	0.063	0.0629999999949	$6.04412663 \times 10^{-11}$	$5.07884013 \times 10^{-12}$
0.4	0.096	0.0959999999950	$6.49500377 \times 10^{-11}$	$5.03508346 \times 10^{-12}$
0.5	0.125	0.1249999999952	$4.92386236 \times 10^{-11}$	$4.82547335 \times 10^{-12}$
0.6	0.144	0.1439999999957	$6.36304374 \times 10^{-11}$	$4.28940217 \times 10^{-12}$
0.7	0.147	0.1469999999965	$4.15211688 \times 10^{-11}$	$3.45659612 \times 10^{-12}$
0.8	0.128	0.1279999999976	$4.98899878 \times 10^{-11}$	$2.43857712 \times 10^{-12}$
0.9	0.081	0.0809999999988	$4.34021510 \times 10^{-11}$	$1.21960775 \times 10^{-12}$
1.0	0	0	0	0

Table 6.
Numerical results for problem 2.

x_i	y_{exact}	y_{CGA}	$\overline{\lvert r_1(i)\rvert}$	$\overline{\lvert e_1(i)\rvert}$
0	0	0	0	0
0.1	0.3090169944	0.3090169942	$8.91393542 \times 10^{-8}$	$1.74298703 \times 10^{-10}$
0.2	0.5877852523	0.5877852478	$1.73533116 \times 10^{-7}$	$4.51602511 \times 10^{-9}$
0.3	0.8090169944	0.8090169921	$2.36294775 \times 10^{-7}$	$2.28544794 \times 10^{-9}$
0.4	0.9510565163	0.9510565072	$1.16235932 \times 10^{-7}$	$9.11584400 \times 10^{-9}$
0.5	1	0.9999999982	$3.12261311 \times 10^{-7}$	$1.75740202 \times 10^{-9}$
0.6	0.9510565163	0.9510565139	$5.42159317 \times 10^{-7}$	$2.39516830 \times 10^{-9}$
0.7	0.8090169944	0.8090169927	$6.54104548 \times 10^{-7}$	$1.72077853 \times 10^{-9}$
0.8	0.5877852523	0.5877852521	$1.03380547 \times 10^{-7}$	$1.99204494 \times 10^{-10}$
0.9	0.3090169944	0.3090169941	$3.45984685 \times 10^{-8}$	$2.87176471 \times 10^{-10}$
1	0	0	0	0

Table 7.
Numerical results of $y_1(x)$ for problem 3.

of CGA. In addition to that, there are many design or analysis problems that are encountered in real life that require solutions which are very accurate regardless of their computational time. Such problems are solved in an offline manner.

6. Convergence analysis

In this section, the effect of different CGA operators and control parameters on the convergence speed—specifically, the average number of generations required for the algorithm to converge to a solution—is analyzed. The analysis is divided into two parts, focusing on the following aspects:

x_i	y_{exact}	y_{CGA}	$\overline{\lvert r_2(i)\rvert}$	$\overline{\lvert e_2(i)\rvert}$
0	1	1	0	0
0.1	1.0050041616	1.0050041681	$3.23108510 \times 10^{-8}$	$6.40738005 \times 10^{-9}$
0.2	1.0200667534	1.0200667556	$2.43406044 \times 10^{-7}$	$2.18877180 \times 10^{-9}$
0.3	1.0453384737	1.0453385141	$4.67771603 \times 10^{-6}$	$4.04596276 \times 10^{-8}$
0.4	1.0810723246	1.0810723718	$6.32416369 \times 10^{-6}$	$4.72321253 \times 10^{-8}$
0.5	1.1276259234	1.1276259652	$7.33519796 \times 10^{-6}$	$4.17938102 \times 10^{-8}$
0.6	1.1854651816	1.1854652182	$7.83019575 \times 10^{-6}$	$3.66241395 \times 10^{-8}$
0.7	1.2551690018	1.2551690056	$5.30712792 \times 10^{-7}$	$3.81259372 \times 10^{-9}$
0.8	1.3374349426	1.3374349463	$1.52175106 \times 10^{-8}$	$3.74691887 \times 10^{-9}$
0.9	1.4330863829	1.4330863854	$3.23108510 \times 10^{-8}$	$2.54984389 \times 10^{-9}$
1	1.5430806348	1.5430806348	0	0

Table 8.
Numerical results of $y_2(x)$ for problem 3.

x_i	y_{exact}	y_{CGA}	$\overline{\lvert r_1(i)\rvert}$	$\overline{\lvert e_1(i)\rvert}$
0	0	0	0	0
0.1	0.3090169944	0.3090169709	$2.46906269 \times 10^{-7}$	$2.34377574 \times 10^{-8}$
0.2	0.5877852523	0.5877852301	$7.33246741 \times 10^{-7}$	$2.21766321 \times 10^{-8}$
0.3	0.8090169944	0.8090169716	$1.31747821 \times 10^{-7}$	$2.27733666 \times 10^{-8}$
0.4	0.9510565163	0.9510564898	$1.57196727 \times 10^{-7}$	$2.64554492 \times 10^{-8}$
0.5	1	0.9999999884	$1.43210365 \times 10^{-7}$	$1.16274315 \times 10^{-8}$
0.6	0.9510565163	0.9510565048	$1.16714975 \times 10^{-7}$	$1.15035971 \times 10^{-8}$
0.7	0.8090169944	0.8090169852	$9.08216851 \times 10^{-7}$	$9.20499440 \times 10^{-9}$
0.8	0.5877852523	0.5877852479	$6.09962319 \times 10^{-7}$	$4.43163132 \times 10^{-9}$
0.9	0.3090169944	0.3090169837	$3.19497157 \times 10^{-7}$	$1.07166703 \times 10^{-8}$
1	0	0	0	0

Table 9.
Numerical results of $y_1(x)$ for problem 4.

1. First Part: evolutionary progress, initialization, and selection.

    a. Evolutionary progress plots of the best-fitness individual. This involves tracking the fitness value of the best individual in each generation over time. These plots help visualize how quickly the algorithm is converging and whether it is getting stuck in local optima.

    b. Effect of various initialization methods. Different initialization methods can have a significant impact on how fast the algorithm converges. The

x_i	y_{exact}	y_{CGA}	$\overline{\lvert r_2(i)\rvert}$	$\overline{\lvert e_2(i)\rvert}$
0	0	0	0	0
0.1	0.09	0.0900000036	$3.55472876 \times 10^{-9}$	$2.82238980 \times 10^{-8}$
0.2	0.16	0.1600000087	$8.68203299 \times 10^{-9}$	$3.31277996 \times 10^{-8}$
0.3	0.21	0.2100000323	$3.22981050 \times 10^{-8}$	$3.94634569 \times 10^{-7}$
0.4	0.24	0.2400000186	$1.85934107 \times 10^{-8}$	$4.50275413 \times 10^{-7}$
0.5	0.25	0.2500000991	$9.91166535 \times 10^{-8}$	$4.94881138 \times 10^{-7}$
0.6	0.24	0.2400000641	$6.41332590 \times 10^{-8}$	$5.20706622 \times 10^{-7}$
0.7	0.21	0.2100000110	$1.10134028 \times 10^{-8}$	$5.45125595 \times 10^{-7}$
0.8	0.16	0.1600000024	$2.36580521 \times 10^{-9}$	$5.35900147 \times 10^{-8}$
0.9	0.09	0.0900000074	$7.42388421 \times 10^{-9}$	$5.38354922 \times 10^{-8}$
1	0	0	0	0

Table 10.
Numerical results of $y_2(x)$ for problem 4.

x_i	y_{exact}	y_{CGA}	$\overline{\lvert r(i)\rvert}$	$\overline{\lvert e(i)\rvert}$
1	17.00000000000	16.99999999996	0	0
1.1	15.75545454545	15.75545454589	2.55851475E-09	4.35460000E-10
1.2	14.77333333333	14.77333333287	2.68958384E-09	4.57768000E-10
1.3	13.99769230769	13.99769230761	4.56979195E-10	7.77780000E-11
1.4	13.38857142857	13.38857142911	3.21303765E-09	5.46860000E-10
1.5	12.91666666666	12.91666666745	4.63763436E-09	7.89326800E-10
1.6	12.56000000000	12.56000000026	1.57377846E-09	2.67857580E-10
1.7	12.30176470588	12.30176470542	2.68257445E-09	4.56575000E-10
1.8	12.12888888888	12.12888888934	2.66423406E-09	4.53453460E-10
1.9	12.03105263157	12.03105263233	4.44627131E-09	7.56756750E-10
2	12.00000000000	11.99999999910	0	0

Table 11.
Numerical results for problem 5.

goal is to see which method provides a better starting point for the population to evolve toward optimal solutions.

c. Effect of common selection schemes. The selection scheme (e.g., tournament selection, roulette wheel selection, and rank-based selection) determines how individuals are chosen for reproduction. Different schemes can balance the exploration and exploitation of the search space, impacting the convergence speed.

x_i	y_{exact}	y_{CGA}	$\overline{\lvert r(i) \rvert}$	$\overline{\lvert e(i) \rvert}$
0.0	$exp\,(1)$	$exp\,(1)$	0	0
0.1	3.0272988228	3.0272988194	$1.63203336 \times 10^{-7}$	$3.44467561 \times 10^{-9}$
0.2	3.3060670808	3.3060670781	$3.96302397 \times 10^{-7}$	$2.62084439 \times 10^{-9}$
0.3	3.5272988228	3.5272988205	$3.23475157 \times 10^{-7}$	$2.34136222 \times 10^{-9}$
0.4	3.6693383448	3.6693383423	$2.94909675 \times 10^{-7}$	$2.43342257 \times 10^{-9}$
0.5	3.7182818285	3.7182818264	$2.71410029 \times 10^{-7}$	$2.09142268 \times 10^{-9}$
0.6	3.6693383448	3.6693383424	$4.97262197 \times 10^{-7}$	$2.32433033 \times 10^{-9}$
0.7	3.5272988228	3.5272988205	$5.84368888 \times 10^{-7}$	$2.34696009 \times 10^{-9}$
0.8	3.3060670808	3.3060670789	$2.97780417 \times 10^{-7}$	$1.86955984 \times 10^{-9}$
0.9	3.0272988228	3.0272988204	$1.98164203 \times 10^{-7}$	$2.41455221 \times 10^{-9}$
1.0	$exp\,(1)$	$exp\,(1)$	0	0

Table 12.
Numerical results for problem 6.

x_i	y_{exact}	y_{CGA}	$\overline{\lvert r_1(i) \rvert}$	$\overline{\lvert e_1(i) \rvert}$
0	0.5	0.5	0	0
0.1	0.59	0.5900000004	$3.92352817 \times 10^{-10}$	$2.88501964 \times 10^{-8}$
0.2	0.66	0.6600000008	$7.52398144 \times 10^{-10}$	$5.51350699 \times 10^{-8}$
0.3	0.71	0.7100000011	$1.06181730 \times 10^{-9}$	$6.49489557 \times 10^{-8}$
0.4	0.74	0.7400000013	$1.25222055 \times 10^{-9}$	$2.38604419 \times 10^{-7}$
0.5	0.75	0.7500000013	$1.30595534 \times 10^{-9}$	$1.12803512 \times 10^{-7}$
0.6	0.74	0.7400000011	$1.09878773 \times 10^{-9}$	$2.15849083 \times 10^{-7}$
0.7	0.71	0.7100000008	$8.34887715 \times 10^{-10}$	$4.23741416 \times 10^{-8}$
0.8	0.66	0.6600000005	$4.80726570 \times 10^{-10}$	$5.21360104 \times 10^{-8}$
0.9	0.59	0.5900000002	$2.40252263 \times 10^{-10}$	$3.69582632 \times 10^{-8}$
1	0.5	0.5	0	0

Table 13.
Numerical results of $y_1(x)$ for problem 7.

2. Second Part: Tuning Parameters.

    a. Rank-based ratio $\{R_{br}\}$. This refers to the selection pressure applied when using rank-based selection. It affects the chances of individuals being selected based on their rank in the population, which can influence convergence rates.

    b. Population size $\{N_p\}$. Larger populations provide more diversity but may slow down convergence, while smaller populations may converge faster but run the risk of premature convergence (getting stuck in local optima).

x_i	y_{exact}	y_{CGA}	$\overline{\lvert r_2(i)\rvert}$	$\overline{\lvert e_2(i)\rvert}$
0	1	1	0	0
0.1	0.9317314234	0.9317314284	$1.28506851 \times 10^{-8}$	$4.99942325 \times 10^{-9}$
0.2	0.8681234454	0.8681234555	$2.19398998 \times 10^{-7}$	$1.00975831 \times 10^{-8}$
0.3	0.8088578935	0.8088579066	$1.11180556 \times 10^{-6}$	$1.30699910 \times 10^{-8}$
0.4	0.7536383164	0.7536383306	$1.20050959 \times 10^{-6}$	$1.41566083 \times 10^{-8}$
0.5	0.7021885013	0.7021885155	$8.62178278 \times 10^{-7}$	$1.41517833 \times 10^{-8}$
0.6	0.6542510919	0.6542511047	$1.06293577 \times 10^{-6}$	$1.28934267 \times 10^{-8}$
0.7	0.6095863011	0.6095863125	$1.46241620 \times 10^{-6}$	$1.14058931 \times 10^{-8}$
0.8	0.5679707120	0.5679707211	$4.29304478 \times 10^{-7}$	$9.06746956 \times 10^{-9}$
0.9	0.5291961600	0.5291961644	$2.72281958 \times 10^{-8}$	$4.48292087 \times 10^{-9}$
1	0.4930686914	0.4930686914	0	0

Table 14.
Numerical results of $y_2(x)$ for problem 7.

x_i	y_{exact}	y_{CGA}	$\overline{\lvert r_1(i)\rvert}$	$\overline{\lvert e_1(i)\rvert}$
0	1	1	0	0
0.1	1.1051709181	1.1051709176	$4.52360174 \times 10^{-10}$	$6.13045170 \times 10^{-9}$
0.2	1.2214027582	1.2214027580	$1.11405697 \times 10^{-10}$	$4.18143742 \times 10^{-9}$
0.3	1.3498588076	1.3498588074	$1.66059423 \times 10^{-10}$	$3.81930265 \times 10^{-9}$
0.4	1.4918246976	1.4918246975	$1.87788104 \times 10^{-10}$	$3.52955665 \times 10^{-9}$
0.5	1.6487212707	1.6487212704	$3.38139288 \times 10^{-10}$	$2.24281038 \times 10^{-9}$
0.6	1.8221188004	1.8221188003	$1.35799902 \times 10^{-10}$	$2.11427986 \times 10^{-9}$
0.7	2.0137527075	2.0137527073	$1.57605628 \times 10^{-10}$	$1.63476344 \times 10^{-9}$
0.8	2.2255409285	2.2255409278	$7.36103139 \times 10^{-10}$	$1.07448717 \times 10^{-9}$
0.9	2.4596031112	2.4596031111	$7.00001307 \times 10^{-10}$	$3.15012016 \times 10^{-9}$
1	2.7182818285	2.7182818285	0	0

Table 15.
Numerical results of $y_1(x)$ for problem 8.

c. Crossover $\{p_{ci}, p_{cc}\}$ and mutation $\{p_{mi}, p_{mc}\}$ probabilities. The crossover probability controls how often crossover (combining parts of two parents to create offspring) occurs, while the mutation probability determines the likelihood of random changes. These probabilities are critical for balancing the algorithm's exploration and exploitation.

d. Maximum nodal residual $\{\sum_{k=1}^{m} r_k(i) \le r_{thr}\}$. This parameter refers to an error tolerance that is allowed at the nodes. A smaller residual means stricter convergence criteria, which can affect how quickly the algorithm converges.

x_i	y_{exact}	y_{CGA}	$\overline{\|r_2(i)\|}$	$\overline{\|e_2(i)\|}$
0	0	0	0	0
0.1	0.1001667500	0.1001667490	$8.02476012 \times 10^{-8}$	$1.06187525 \times 10^{-9}$
0.2	0.2013360025	0.2013360002	$3.96756814 \times 10^{-8}$	$2.37455608 \times 10^{-9}$
0.3	0.3045202934	0.3045202917	$4.46209980 \times 10^{-8}$	$1.78401083 \times 10^{-9}$
0.4	0.4107523258	0.4107523239	$5.46656354 \times 10^{-8}$	$1.93059550 \times 10^{-9}$
0.5	0.5210953055	0.5210953042	$3.33396211 \times 10^{-8}$	$1.28153001 \times 10^{-9}$
0.6	0.6366535821	0.6366535804	$3.02547742 \times 10^{-8}$	$1.76460065 \times 10^{-9}$
0.7	0.7585837018	0.7585836993	$3.84077958 \times 10^{-8}$	$2.54339745 \times 10^{-9}$
0.8	0.8881059822	0.8881059811	$2.51190269 \times 10^{-8}$	$1.11469865 \times 10^{-9}$
0.9	1.0265167257	1.0265167250	$2.78001644 \times 10^{-9}$	$6.74489897 \times 10^{-10}$
1	1.1752011936	1.1752011936	0	0

Table 16.
Numerical results of $y_2(x)$ for problem 8.

x_i	HA method	RKHS method	RPS method	CGA
0	0	0	0	0
0.1	$2.35949582 \times 10^{-3}$	$2.19034993 \times 10^{-3}$	$8.09988278 \times 10^{-8}$	$1.74298703 \times 10^{-10}$
0.2	$4.74045574 \times 10^{-3}$	$2.80355432 \times 10^{-3}$	$1.62722147 \times 10^{-7}$	$4.51602511 \times 10^{-9}$
0.3	$7.15177665 \times 10^{-3}$	$2.24628757 \times 10^{-3}$	$2.45620439 \times 10^{-7}$	$2.28544794 \times 10^{-9}$
0.4	$9.56258009 \times 10^{-3}$	$1.58473337 \times 10^{-3}$	$3.29652934 \times 10^{-7}$	$9.11584400 \times 10^{-9}$
0.5	$1.18508499 \times 10^{-2}$	$1.40577968 \times 10^{-3}$	$4.14121107 \times 10^{-7}$	$1.75740202 \times 10^{-9}$
0.6	$1.37257579 \times 10^{-2}$	$1.57048642 \times 10^{-3}$	$4.97487119 \times 10^{-7}$	$2.39516830 \times 10^{-9}$
0.7	$1.46315590 \times 10^{-2}$	$1.62831184 \times 10^{-3}$	$5.76188707 \times 10^{-7}$	$1.72077853 \times 10^{-9}$
0.8	$1.36485833 \times 10^{-2}$	$1.32090548 \times 10^{-3}$	$6.35055770 \times 10^{-7}$	$1.99204494 \times 10^{-10}$
0.9	$9.40524600 \times 10^{-3}$	$7.18291366 \times 10^{-4}$	$5.88955323 \times 10^{-7}$	$2.87176471 \times 10^{-10}$
1	0	0	0	0

Table 17.
Numerical comparison for the average absolute error $\overline{(|e_1(i)|)}$ of $y_1(x)$ for Problem 3.

e. Step size effect $\{h\}$. Step size relates to the number of unknown nodes within the boundary value problem according to Eq. (3). To some extent, smaller step size leads to better accuracy at the expense of slower down convergence.

This analysis help understand how tweaking various components of the CGA affects its overall performance, especially in terms of convergence speed and solution quality.

Throughout this convergence analysis section, two problems out of the eight problems are selected for performing this task without losing the generality of the performed analysis. This fact is supported from the results obtained in **Table 4**

x_i	HA method	RKHS method	RPS method	CGA
0	0	0	0	0
0.1	$2.63024554 \times 10^{-3}$	$1.87526124 \times 10^{-2}$	$3.71983333 \times 10^{-8}$	$6.40738005 \times 10^{-9}$
0.2	$5.28040299 \times 10^{-3}$	$2.29239231 \times 10^{-2}$	$6.96598341 \times 10^{-8}$	$2.18877180 \times 10^{-9}$
0.3	$7.91142058 \times 10^{-3}$	$1.92964992 \times 10^{-2}$	$9.65249105 \times 10^{-8}$	$4.04596276 \times 10^{-8}$
0.4	$1.04098574 \times 10^{-2}$	$1.27708646 \times 10^{-2}$	$1.16636828 \times 10^{-7}$	$4.72321253 \times 10^{-8}$
0.5	$1.25405615 \times 10^{-2}$	$6.53617494 \times 10^{-3}$	$1.28558841 \times 10^{-7}$	$4.17988102 \times 10^{-8}$
0.6	$1.39107975 \times 10^{-2}$	$2.20945555 \times 10^{-3}$	$1.30605800 \times 10^{-7}$	$3.66241395 \times 10^{-8}$
0.7	$1.39677744 \times 10^{-2}$	$7.18511912 \times 10^{-5}$	$1.20885677 \times 10^{-7}$	$3.81259372 \times 10^{-9}$
0.8	$1.20521694 \times 10^{-2}$	$4.74726833 \times 10^{-4}$	$9.73443477 \times 10^{-8}$	$3.74691887 \times 10^{-9}$
0.9	$7.52731717 \times 10^{-3}$	$2.94339175 \times 10^{-4}$	$5.78058714 \times 10^{-8}$	$2.54984389 \times 10^{-9}$
1	0	0	0	0

Table 18.
Numerical comparison for the average absolute error $\overline{(|e_2(i)|)}$ *of* $y_2(x)$ *for problem 3.*

x_i	HA method	RKHS method	RPS method	CGA
0	Failed	0	0	0
0.1	Failed	$2.56831516 \times 10^{-4}$	$9.59760145 \times 10^{-7}$	$3.92352817 \times 10^{-10}$
0.2	Failed	$3.84855823 \times 10^{-4}$	$2.09206207 \times 10^{-6}$	$7.52398144 \times 10^{-10}$
0.3	Failed	$3.88834896 \times 10^{-4}$	$3.39993180 \times 10^{-6}$	$1.06181730 \times 10^{-9}$
0.4	Failed	$3.39418875 \times 10^{-4}$	$4.81835117 \times 10^{-6}$	$1.25222055 \times 10^{-9}$
0.5	Failed	$2.73565020 \times 10^{-4}$	$6.20062360 \times 10^{-6}$	$1.30595534 \times 10^{-9}$
0.6	Failed	$2.07350748 \times 10^{-4}$	$7.30473987 \times 10^{-6}$	$1.09878773 \times 10^{-9}$
0.7	Failed	$1.46368749 \times 10^{-4}$	$7.77974390 \times 10^{-6}$	$8.34887715 \times 10^{-10}$
0.8	Failed	$9.17097885 \times 10^{-5}$	$7.15209851 \times 10^{-6}$	$4.80726570 \times 10^{-10}$
0.9	Failed	$4.29160977 \times 10^{-5}$	$4.81205121 \times 10^{-6}$	$2.40252263 \times 10^{-10}$
1	Failed	0	0	0

Table 19.
Numerical comparison for the average absolute error $\overline{(|e_1(i)|)}$ *of* $y_1(x)$ *for problem 7.*

showing the convergence data of the eight boundary value problems where the convergence data is, to some extent, insensitive to the selected problem. The two problems are problem 3 representing a linear nonsingular system and problem 7 representing a nonlinear nonsingular system of equations as given in **Table 1** [32].

First, three types of initialization methods are explored which include using MNGF solely, MTHF solely, and the mixed-type of MNGF and MTHF with a 50–50% application ratio. It is observed from the results shown in **Table 22** that the initialization methods have minor effect on the convergence speed. This is justified as the initial population expires after few tens of generations. Once this happens, the convergence

x_i	HA method	RKHS method	RPS method	CGA
0	Failed	0	0	0
0.1	Failed	$5.02938361 \times 10^{-4}$	$9.77486288 \times 10^{-6}$	$4.99942325 \times 10^{-9}$
0.2	Failed	$7.99478237 \times 10^{-4}$	$2.03608577 \times 10^{-5}$	$1.00975831 \times 10^{-8}$
0.3	Failed	$9.35309345 \times 10^{-4}$	$3.17932014 \times 10^{-5}$	$1.30699910 \times 10^{-8}$
0.4	Failed	$9.48609720 \times 10^{-4}$	$4.38820180 \times 10^{-5}$	$1.41566083 \times 10^{-8}$
0.5	Failed	$8.74060480 \times 10^{-4}$	$5.59315633 \times 10^{-5}$	$1.41517833 \times 10^{-8}$
0.6	Failed	$7.41537184 \times 10^{-4}$	$6.63582152 \times 10^{-5}$	$1.28934267 \times 10^{-8}$
0.7	Failed	$5.74700989 \times 10^{-4}$	$7.22141423 \times 10^{-5}$	$1.14058931 \times 10^{-8}$
0.8	Failed	$3.90212246 \times 10^{-4}$	$6.86230881 \times 10^{-5}$	$9.06746956 \times 10^{-9}$
0.9	Failed	$1.97622938 \times 10^{-4}$	$4.81342728 \times 10^{-5}$	$4.48232087 \times 10^{-9}$
1	Failed	0	0	0

Table 20.
Numerical comparison for the average absolute error $\overline{(|e_2(i)|)}$ *of* $y_2(x)$ *for problem 7.*

Problem number	HA method	RKHS method	RPS method	CGA
3	$9.62815838 \times 10^{-3}$	$5.48884150 \times 10^{-3}$	$2.43667934 \times 10^{-7}$	$1.15150865 \times 10^{-8}$
7	Failed	$4.49795612 \times 10^{-4}$	$2.56439769 \times 10^{-5}$	$5.65247209 \times 10^{-9}$

Table 21.
Average absolute error $\overline{(|e(i)|)}$ *for all nodes for all variables of problems 3 and 7.*

Initialization method	Problem 3	Problem 7
MNGF	***997***	1307
MTHF	***1135***	1295
Mixed type	1084	*1256*

Table 22.
Effect of initialization schemes on the convergence speed of CGA.

speed becomes primarily governed by the selection mechanism, crossover operator, and mutation operator. The best convergence speed is highlighted and underlined. MNGF results in the fastest convergence speed for problem 3, while mixed type initialization and MTHF results in the fastest convergence speed for problem 7. However, since the solution of any given problem is assumed to be unknown, it is better to have a diverse initial population by using of the mixed-type initialization method. Based on that, mixed-type initialization method is used as the algorithm default method.

After that, the effect of the most frequently used selection scenarios of the performance on the CGA is explored. Selection scenarios impact how individuals are chosen for reproduction and how diversity and convergence are balanced. Six selection scenarios are used in the algorithm, which include rank-based, tournament with replacement, tournament without replacement, roulette wheel, stochastic universal, and half biased selection [29, 31, 39, 40].

1. Rank-based selection. According to this mechanism, individuals are ranked based on fitness, and the top portion (determined by the rank-based ratio, R_{br},) is selected to form a mating subpopulation of the size $\{R_{br}N_p\}$. After that, parents are randomly chosen from this mating subpopulation. This method ensures that only the best individuals have a chance to reproduce, which increases selection pressure, potentially leading to faster convergence.

2. Tournament selection scheme. This mechanism selects a pair of individuals randomly from the parent population, and the one with higher fitness is placed in the mating pool. Tournament selection has two types depending on whether the pair of selected parents are returned into the parent population or not.

    a. Tournament without replacement. Two parents are randomly selected from the population. Then, the individual with the higher fitness (the better individual) is chosen and added to the mating pool. After that, both selected individuals are set aside and are not returned to the population until all other individuals have been removed. Next, when half of the mating pool is filled, the process is repeated for a second round in order to fill the mating pool.

    b. Tournament with replacement. In this mechanism, the same steps of that without replacement are applied with the exception that both of the selected parents are returned into the original population for the next selection operation until the mating pool is full.

    It is to be noted that tournament without replacement tends to maintain higher diversity in the mating pool but may slow down convergence, while tournament with replacement allows for faster convergence as better individuals are more likely to be selected multiple times, but it suffers the risk of premature convergence due to reduced diversity.

3. Roulette wheel selection. This selection process is fitness proportionality scheme that is analogous to a spinning roulette wheel, where each individual in the population is assigned a segment (slot) on the wheel. The size of each slot is proportional to the fitness of the individual. This means that individuals with higher fitness have larger slots, giving them a greater probability of being selected. This process is repeated until the mating pool is filled with selected individuals.

4. Stochastic universal selection. This method is an enhanced version of roulette wheel selection that addresses some of the limitations of the basic roulette wheel approach, particularly in ensuring more consistent proportional selection. Instead of spinning the wheel multiple times as in roulette wheel selection, N_p equidistant markers are placed around the wheel. After that, s single "spin" of the wheel is performed, but instead of stopping once to select one individual, the markers simultaneously select multiple individuals at once based on their positions relative to the fitness slots. The number of markers that fall inside a

particular individual's slot determines the number of copies (or offspring) that the individual will have in the next generation.

5. Half-biased selection. According to this mechanism, one mate is selected as in the roulette wheel selection, while the other mate is randomly selected from the original population without any attention to its fitness.

The convergence information for the six selection schemes is given in **Table 23** where the best convergence speed is highlighted and underlined. From this table, it is observed that the fastest convergence speed is achieved using the rank-based selection scheme. The second place is for the tournament selection approaches with nearly comparable speeds. The fitness proportionate methods including roulette wheel, stochastic universal, and half-biased selection schemes have slower convergence speed with half-biased selection method coming at the bottom.

The effect of the rank-based ratio, R_{br}, is studied next. The used values for the rank-based ratio lie within the range $[0.1,1]$ with a step size of 0.1. The best convergence speed in **Table 24** is highlighted and underlined. It can be easily observed that the best convergence speed for the two problems is achieved when $R_{br} = 10\%$. Fur-

Selection method	Problem 3	Problem 7
Rank-based	***1084***	***1256***
Tournament with replacement	1155	1291
Tournament without replacement	1121	1270
Roulette wheel	1436	1659
Stochastic universal	1462	1699
Half-biased	1516	1704

Table 23.
Convergence speed of CGA using different selection schemes.

R_{br}	Problem 3	Problem 7
0.1	***1084***	***1256***
0.2	1139	1344
0.3	1186	1383
0.4	1235	1429
0.5	1302	1496
0.6	1343	1554
0.7	1381	1575
0.8	1472	1662
0.9	1523	1711
1.0	1619	1847

Table 24.
Effect of the rank-based ratio on the convergence speed of CGA.

thermore, it is observed that the CGA suffers from slower convergence speed as the rank-based ratio increases.

Figure 5 shows the evolutionary progress plots of the best-of-generation individual of the third and seventh problems. It is clear from the figure that the convergence curve can be divided into two stages:

1. Coarse-tuning stage. This represents the initial convergence stage where the best fitness approaches to unity value (say 0.99) within about $\{20\% - 25\%\}$ of the generations required for convergence. CGA converges to the near optimal solution very fast in this stage.

2. Fine-tuning stage. This represents the final convergence stage where the best-fitness value changes from about 0.99 until the convergence criteria is met (about 0.99999). This stage takes about $\{75\% - 80\%\}$ of the generations required for convergence. CGA converges to the optimal solution very slow in this stage.

After that, the influence of the population size is studied as shown in **Table 25** for the third problem. The population size is increased starting with 100 individuals and ending with 1000 individuals with a step size of 100 individuals. It is observed that small population sizes reduce the number of fitness evaluations within each generation, which means lower execution times, but this comes at the cost of convergence quality since the algorithm requires more generations to converge to an optimal solution. Additionally, the algorithm is more prone to being trapped in local minima, as there is less genetic diversity to explore the solution space effectively. On the other hand, as the population size increases (up to 1000 individuals), the number of fitness evaluations grows, leading to longer execution times. However, larger populations help improve the convergence speed because they maintain more genetic diversity, which allows for better exploration of the solution space and reduces the risk of local optima.

From **Table 25**, it can be also observed that there is a point of diminishing returns where after reaching a population size of about 500 individuals, the improvement in convergence speed becomes almost negligible, meaning that increasing the population size beyond this point does not significantly speed up convergence. This means that

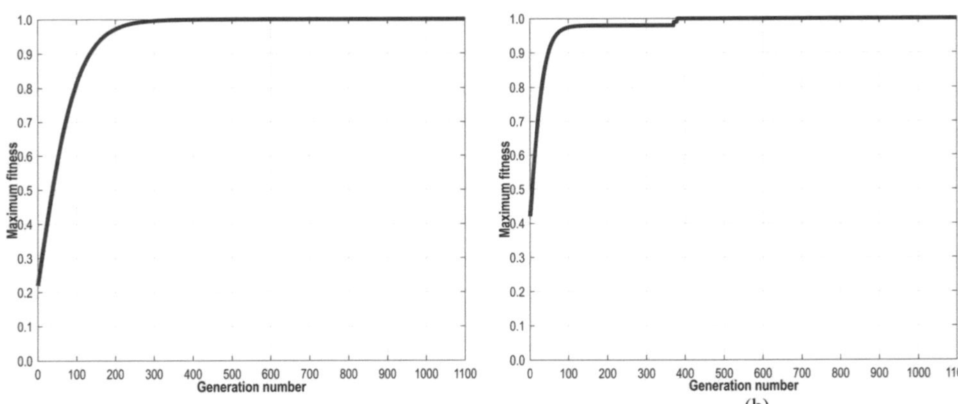

Figure 5.
Evolutionary progress plots for the best-of-generation individual for (a) problem 3 and (b) problem 7.

N_p	Average number of generations	Average fitness	Average absolute residual	Average absolute error
100	1794	0.99985367	$2.24395284 \times 10^{-3}$	$1.05800717 \times 10^{-5}$
200	1584	0.99994801	$1.35985578 \times 10^{-4}$	$8.84899864 \times 10^{-6}$
300	1360	0.99997652	$2.06965887 \times 10^{-5}$	$6.55745545 \times 10^{-7}$
400	1136	0.99999506	$8.84654342 \times 10^{-6}$	$9.89660807 \times 10^{-8}$
500	1084	0.99999868	$1.62682994 \times 10^{-6}$	$1.15150865 \times 10^{-8}$
600	961	0.99999887	$7.41344141 \times 10^{-7}$	$8.00965709 \times 10^{-9}$
700	944	0.99999896	$3.68834042 \times 10^{-7}$	$4.39225426 \times 10^{-9}$
800	918	0.99999900	$1.05473607 \times 10^{-7}$	$1.69075768 \times 10^{-9}$
900	893	0.99999900	$8.97735465 \times 10^{-8}$	$9.89521725 \times 10^{-10}$
1000	875	0.99999900	$2.81066591 \times 10^{-8}$	$7.41876214 \times 10^{-10}$

Table 25.
Effect of the population size on the convergence speed, the average fitness, and the corresponding errors of CGA for problem 3.

(p_{mc}, p_{cc})	0.1	0.3	0.5	0.7	0.9
0.1	1520	1494	1181	1414	1574
0.3	1461	1293	1120	1314	1462
0.5	1349	1118	***1084***	1145	1280
0.7	1433	1312	1169	1309	1325
0.9	1658	1450	1218	1433	1536

Table 26.
Effect of the curve crossover probability and the curve mutation probability on the convergence speed of CGA for problem 3.

the population size should be at least of about 50 times the number of unknown nodes in the problem in order to obtain best results. The proper selection of the population size clearly implies the importance of the *trade-off between execution time and accuracy*.

An investigation of the crossover and mutation probabilities is explored next. These probabilities are problem-dependent and have to be determined by simulations. They play a vital role in the efficiency of the algorithm. **Table 26** shows the collective influence of the curve crossover probability $\{p_{cc}\}$ and the curve mutation probability $\{p_{mc}\}$ on the convergence speed of the algorithm for the third problem, while their collective influence on the average fitness for the seventh problem is shown in **Table 27**. The probability values are increased from 0.1 till 0.9 with a step size of 0.2 for both p_{cc} and p_{mc}, while the individual crossover probability $\{p_{ci}\}$ and the individual mutation probability $\{p_{mi}\}$ are kept at 0.9. Based on the results shown in **Tables 26** and **27**, it is clear the best performance of the algorithm is achieved at the center when for the optimal values $(p_{mc}, p_{cc}) = (0.5, 0.5)$. Furthermore, when p_{cc} and p_{mc} are deviated from these optimal values in an increasing or decreasing manner,

(p_{mc}, p_{cc})	0.1	0.3	0.5	0.7	0.9
0.1	0.99517043	0.99951767	0.99997410	0.99904693	0.99170643
0.3	0.99906886	0.99997806	0.99999287	0.99998633	0.99952671
0.5	0.99993423	0.99999679	***0.99999900***	0.99999507	0.99994948
0.7	0.99940430	0.99998242	0.99999852	0.99990694	0.99962965
0.9	0.99800981	0.99985511	0.99990967	0.99949034	0.99422627

Table 27.
Effect of the curve crossover probability and the curve mutation probability on the average fitness of CGA for problem 7.

Number of nodes	Average number of generations	Average absolute residual	Average absolute error
10	1256	$9.41844056 \times 10^{-8}$	$8.24377603 \times 10^{-10}$
20	1583	$3.53945884 \times 10^{-9}$	$7.32416369 \times 10^{-11}$
40	1859	$2.59147738 \times 10^{-10}$	$6.43406044 \times 10^{-12}$
80	2101	$1.81259371 \times 10^{-11}$	$6.75384829 \times 10^{-13}$

Table 28.
Effect of the number of nodes on the convergence speed and the corresponding errors of $y_1(x)$ for problem 7.

Number of nodes	Average number of generations	Average absolute residual	Average absolute error
10	1256	$7.09847528 \times 10^{-7}$	$1.04805666 \times 10^{-8}$
20	1583	$1.19273048 \times 10^{-7}$	$4.18233471 \times 10^{-9}$
40	1859	$9.25765710 \times 10^{-8}$	$5.28391737 \times 10^{-10}$
80	2101	$5.81643771 \times 10^{-9}$	$3.90307129 \times 10^{-11}$

Table 29.
Effect of the number of nodes on the convergence speed and the corresponding errors of $y_2(x)$ for problem 7.

then the average number of generations required for convergence increases, while the average fitness is decreases. Additionally, p_{cc} and p_{mc} have a minor effect on the performance of the CGA. These results are logical since too high crossover probability can lead to a lack of diversity, while too low crossover probability may result in inadequate exploration of the solution space. Likewise, a mutation probability that is too high may introduce too much randomness, hindering convergence, while too low mutation probability may lead to stagnation. These optimal values of $(p_{mc}, p_{cc}) = (0.5, 0.5)$ are set as the algorithm default values.

The effect of the number of unknown nodes of the boundary value problem on the convergence speed and the corresponding errors of CGA is studied next. The number of unknown nodes is within the range $(10, 80)$ as given in **Tables 28** and **29** for the seventh problem. It is observed that an increase in the number of unknown nodes

Maximum nodal residual	Average number of generations	Average fitness	Average absolute residual	Average absolute error
0.1	141	0.83424743	$4.84328825 \times 10^{-2}$	$7.27919815 \times 10^{-3}$
0.01	358	0.92732053	$9.35421009 \times 10^{-3}$	$1.28834055 \times 10^{-3}$
0.001	796	0.99252428	$1.08874528 \times 10^{-3}$	$1.32204652 \times 10^{-4}$
0.0001	995	0.99980020	$8.11219421 \times 10^{-4}$	$1.42277479 \times 10^{-5}$
0.00001	1220	0.99985771	$7.91614021 \times 10^{-5}$	$1.13731561 \times 10^{-6}$
0.000001	1407	0.99995942	$2.25519919 \times 10^{-5}$	$2.31464404 \times 10^{-7}$
0.0000001	1534	0.99999149	$8.38288398 \times 10^{-6}$	$3.09797357 \times 10^{-8}$
0.00000001	1592	0.99999624	$4.14596602 \times 10^{-7}$	$5.90831295 \times 10^{-9}$
0.000000001	1602	0.99999683	$1.04334938 \times 10^{-7}$	$1.81022863 \times 10^{-9}$
0.0000000001	1661	0.99999735	$9.13754001 \times 10^{-8}$	$7.86811475 \times 10^{-10}$

Table 30.
Influence of the maximum nodal residual on the convergence speed, average fitness, and the corresponding errors of CGA for problem 3.

results in a reduction in the average absolute error leading to an enhancement in the accuracy of the solutions. However, increasing the number of unknown nodes increases the number of generations required for convergence.

Finally, an investigation of the maximum nodal residual of the best-of-generation individual on the convergence speed, the average fitness, and the corresponding errors is explored. The maximum nodal residual is set within the range $(10^{-1}, 10^{-10})$ with a multiplication factor of 10^{-1} as given in **Table 30** for the third problem. It is observed that, as the maximum nodal residual decreases, the number of generations required for convergence increases, while the average absolute error decreases.

7. Conclusions

In this chapter, second-order ordinary boundary value problems were solved using continuous genetic algorithms. CGA is a variant of genetic algorithms that uses curves for representing the unknown nodal values throughout the evolution process. It is applied when the unknown variables of the optimization problem are coupled or correlated with each other which is the case of the solutions obtained for the ordinary boundary value problems. Based on that, CGA is a perfect optimization tool for such problems.

Central to the CGA approach is the representation of each derivative in the ordinary boundary value problem by its finite difference approximation using the proper form of the finite difference formulas (forward, backward, hybrid, or central) and the required number of data points according to the desired accuracy. After that, the overall residue for all unknown nodes in the given problem is determined where the solution to the boundary value problem is finally converted into maximization of the fitness function.

All classifications of ordinary boundary value problems were covered in this work which include eight general cases covering the combinations of linearity (linear/

nonlinear), singularity (singular/nonsingular), and number of equations (single equation /system of equations). On the contrary to other numerical or analytical methods for the solution of such problems, no modifications were performed in the CGA while solving all of these classifications. This fact points to the robustness and simplicity of the algorithm where mathematicians, engineers, and scientists can use the one and only one numerical solver which solves all encountered cases in real life. The numerical solutions obtained using CGA are of excellent accuracy where the average absolute error for the eight selected problems is of the order of 10^{-9} within a range of $\{10^{-8}, 10^{-12}\}$. This proves the three-pillar facts about CGAs related to simplicity, robustness, and accuracy.

The solution methodology is then compared with other well-known numerical methods for the solution of the given ordinary boundary value problems including residual power series method, reproducing kernel Hilbert space method, and the homotopy analysis method. It was observed that the solutions obtained using CGA are much better than that of these methods in terms of the accuracy.

The convergence analysis of CGA was then performed for selected operators of CGA and its controlling parameters including the initialization methods, selection methods, rank-based ratios, population size, and the crossover and mutation probabilities. It was observed that all of these genetic-related parameters and operators have a minor effect on the convergence speed of the algorithm pointing to an additional advantage of its relative insensitivity to operators and controlling parameters changes. The analysis includes also the effect of the step size and the maximum nodal residue on the convergence speed. It was observed that these two problem-related parameters have major effect on the convergence speed of the algorithm.

Author details

Zaer Salem Abo-Hammour
Department of Mechatronics Engineering, School of Engineering, The University of Jordan, Amman, Jordan

*Address all correspondence to: zaer_hr@yahoo.com; zaer@ju.edu.jo

References

[1] Iqbal M, Zainuddin N, Daud H, Kanan R, Soomro H, Jusoh R, et al. A modified basis of cubic B-spline with free parameter for linear second order boundary value problems: Application to engineering problems. Journal of King Saud University Science. 2024;**36**(9). Available from: https://www.sciencedirect.com/science/article/pii/S1018364724003094. DOI: 10.1016/j.jksus.2024.103397

[2] Gurbuz B, Sezer M. Laguerre polynomial solutions of a class of initial and boundary value problems arising in science and engineering fields. Acta Physica Polonica A. 2016;**130**(1): 194-197

[3] Simos TE, Papakaliatakis G. Modified Runge-Kutta Verner methods for the numerical solution of initial and boundary-value problems with engineering applications. Applied Mathematical Modelling. 1998;**22**(9):667-670

[4] Feng HY, Yue XK, Wang XC. A class of linearization-based collocation methods for initial value and boundary value engineering problems. Computer Physics Communications. 2023:283. DOI: 10.1016/j.cpc.2022.108601. Available from: https://www.sciencedirect.com/science/article/abs/pii/S0010465522003204?via%3Dihub

[5] Gasparin S, Berger J, Dutykh D, Mendes N. An adaptive simulation of nonlinear heat and moisture transfer as a boundary value problem. International Journal of Thermal Sciences. 2018;**133**: 120-139

[6] Khan Y. A series solution of the boundary value problem arising in the application of fluid mechanics. International Journal of Numerical Methods for Heat & Fluid Flow. 2018; **28**(10):2480-2490

[7] Xenophontos C, Oberbroeckling L. A numerical study on the finite element solution of singularly perturbed systems of reaction-diffusion problems. Applied Mathematics and Computation. 2007; **187**(2):1351-1367

[8] Bellew S, Oriordan E. A parameter robust numerical method for a system of two singularly perturbed convection-diffusion equations. Applied Numerical Mathematics. 2004;**51**:171-186

[9] Gupta Y, Kumar M. A computational approach for solution of singular boundary value problem with applications in human physiology. National Academy Science Letters-India. 2012;**35**(3):189-193

[10] Caglar H, Caglar N, Ozer M. B-spline solution of non-linear singular boundary value problems arising in physiology. Chaos, Solitons and Fractals. 2009;**39**(3): 1232-1237

[11] Hiltman P, Lory P. On oxygen diffusion in a spherical cell with Michaelis-Menten oxygen uptake kinetics. Bulletin of Mathematical Biology. 1983;**45**(5):661-664

[12] Konyukhova NB, Lima PM, Morgado ML, Soloviev MB. Bubbles and droplets in nonlinear physics models: Analysis and numerical simulation of singular nonlinear boundary value problem. Computational Mathematics and Mathematical Physics. 2008;**48**(11):2018-2058

[13] Ebaid A, Wazwaz AM, Alali E, Masaedeh BS. Hypergeometric series solution to a class of second-order boundary value problems via Laplace transform with applications to Nanofluids. Communications in Theoretical Physics. 2017;**67**(3):231-234

[14] Avvakumov S, Kiselev Y. Boundary value problem for ordinary differential equations with applications to optimal control. In: Eighth World Multi-Conference on Systemics, Cybernetics and Informatics, Vol XIV, Proceedings: Computer and Information Systems, Technologies and Applications. Orlando, FL, USA: International Institute of Informatics and Systemics; 2004. pp. 156-160

[15] Prazak P. Shooting method for boundary value problems of ordinary differential equations in economics. In: 40th International Conference on Mathematical Methods in Economics. Jihlava: Czech Republic; 2022. pp. 300-305

[16] Liao S. On the homotopy analysis method for nonlinear problems. Applied Mathematics and Computation. 2004; **147**(2):499-513

[17] Abu Arqub O, AL-Smadi M, Momani S, Hayat T. Numerical solutions of fuzzy differential equations using reproducing kernel Hilbert space method. Soft Computing. 2016;**20**(8): 3283-3302

[18] Dehghan M, Saadatmandi A. The numerical solution of a nonlinear system of second-order boundary value problems using the sinc-collocation method. Mathematical and Computer Modeling. 2007;**46**(11–12):1434-1441

[19] Abu AO. Application of residual power series method for the solution of time-fractional Schrodinger equations in one-dimensional space. Fundamenta Informaticae. 2019;**166**(2):87-110

[20] Geng F, Cui MG. Solving a nonlinear system of second order boundary value problems. Journal of Mathematical Analysis and Applications. 2007;**327**(2): 1167-1181

[21] Gnativ LB, Kutniv MV, Makarov VL. Generalized three-point difference schemes of high-order accuracy for systems of second order nonlinear ordinary differential equations. Differential Equations. 2009;**45**(7): 998-1019

[22] Matthews S, Oriordan E, Shishkin GI. A numerical method for a system of singularly perturbed reaction–diffusion equations. Journal of Computational and Applied Mathematics. 2002;**145**(1):151-166

[23] Keller HB. Numerical Methods for Two-Point Boundary-Value Problems. NY, USA: Dover Publications; 2018

[24] Burden RL, Faires JD, Burden AM. Numerical Analysis. 10th ed. Boston, MA, USA: Cengage Learning; 2015

[25] Chapra S, Canale R. Numerical Methods for Engineers. 8th ed. NY, USA: McGraw-Hill Education; 2020

[26] Rao SS. The Finite Element Method in Engineering. 5th ed. UK: Butterworth-Heinemann; 2010

[27] Goldberg DE. Genetic Algorithms in Search, Optimization and Machine Learning. WA, USA: Addison-Wesley; 1989

[28] Samii YR, Michielssen E. Electromagnetic Optimization by Genetic Algorithms. NY, USA: Wiley-Interscience; 1999

[29] Abo-Hammour ZS. Advanced Continuous Genetic Algorithms and their Applications in the Motion Planning of Robotic Manipulators and the Numerical Solution of Boundary Value Problems. Pakistan: Quiad-Azam University; Islamabad; 2002

[30] Abo-Hammour ZS, Yusuf M, Mirza NM, Mirza SM. Numerical solution of second-order, two-point boundary value problems using continuous genetic algorithms. International Journal for Numerical Methods in Engineering. 2004;**61**(8):1219-1242

[31] Abu Arqub O, Abo-Hammour Z. Numerical solution of systems of second-order boundary value problems using continuous genetic algorithm. Information Sciences. 2014;**279**:396-415

[32] Abu Arqub O, Abo-Hammour Z, Momani S. Application of continuous genetic algorithm for nonlinear system of second-order boundary value problems. Applied Mathematics and Information Sciences. 2014;**8**(1):235-248

[33] Abo-Hammour Z, Abu Arqub O, Momani S, Shawagfeh N. Optimization solution of Troesch's and Bratu's problems of ordinary type using novel continuous genetic algorithm. Discrete Dynamics in Nature and Society. 2014;**2014**:25. DOI: 10.1155/2014/401696. Article ID 401696

[34] Abu, Arqub O, Abo-Hammour Z, Momani S. Solving singular two-point boundary value problems using continuous genetic algorithm. Abstract and Applied Analysis. 2012:**2012**:25. Article ID 205391. DOI: 10.1155/2012/205391

[35] Abo-Hammour Z, Abu Arqub O, Alsmadi O, Momani S. An optimization algorithm for solving Systems of Singular Boundary Value Problems. Applied Mathematics and Information Sciences. 2014;**8**(6):2809-2821

[36] Abo-Hammour ZS, Samhouri AD, Mubarak Y. Continuous genetic algorithm as a novel solver for stokes and nonlinear Navier stokes problems. Mathematical Problems in Engineering. 2014;**2014**:18. Article ID 649630. DOI: 10.1155/2014/649630

[37] Albadarneh RB, Abo-Hammour Z, Alsmadi O, Shawagfeh N. A novel continuous genetic algorithm technique for the solution of partial differential equations. Italian Journal of Pure and Applied Mathematics. 2021;**45**:216-236

[38] Abo-Hammour ZS, Albadarneh RB, Saraireh MS. Solution of Laplace equation using continuous genetic algorithms. Kuwait Journal of Science & Engineering. 2010;**37**(2A):1-15

[39] Abo-Hammour ZS, Mirza NM, Mirza SM. Cartesian path generation of robot manipulators using continuous genetic algorithms. Robotics and Autonomous Systems. 2002;**41**(4):179-223

[40] Abo Hammour ZS. A novel continuous genetic algorithm for the solution of the cartesian path generation problem of robot manipulators. In: Lui JX, editor. Robot Manipulators: New Research. UK: Nova Publishers; 2005. pp. 133-190

[41] Abo-Hammour ZS, Alsmadi OMK, Bataineh SI. Continuous genetic algorithms for collision-free Cartesian path planning of robot manipulators. International Journal of Advanced Robotic Systems. 2011;**8**(6):14-36

[42] Momani S, Abo-Hammour ZS, Alsmadi OMK. Solution of inverse kinematics problem using genetic algorithms. Applied Mathematics and Information Science. 2016;**10**(1):225-233

[43] Abo-Hammour ZS, Asasfeh AG, Al-Smadi AM. A novel continuous genetic algorithm for the solution of optimal control problems. Optimal Control

Applications & Methods. 2011;**32**(4): 414-432

[44] Alsmadi OMK, Abo-Hammour ZS, Al-Smadi AMA. Robust and efficient genetic algorithm for solving a chemical reactor problem: Theory, application and convergence analysis. Transactions of the Institute of Measurement and Control. 2012;**34**(5):594-603

[45] Li J. General explicit difference formulas for numerical differentiation. Journal of Computational and Applied Mathematics. 2005;**183**(1):29-52

Chapter 3

Perspective Chapter: Pseudo-Differential Operators on $\mathbb{Z}^N$

Perrin G. Kibiti Pembe and Linda N.A. Botchway

Abstract

Referring to the work published by Shahla Molahajloo, this chapter extends the dimension framework from his study of the space $\mathbb{Z}$ to a more general setting. The focus of the chapter is on pseudo-differential operators in the lattice $\mathbb{Z}^N$, examining the key properties and characterization. This will help us establish sufficient conditions for the L^p-boundedness and L^p-compactness of these operators. In addition to that, the chapter will leverage the Fourier transform as a tool to address the broader problem. The primary goal of this chapter is to improve the understanding of how to compute the inverse of nonlinear differential operators.

Keywords: pseudo-differential operators, Fourier transform, Stwartz space, Hilbert-Schmidt operators, L^p-boundedness, L^p-compactness

1. Introduction

Pseudo-differential operators (ΨDOs) appeared around 1950 due to the need of finding parametrix for elliptic operators, which is considered an approximation to a solution of elliptic equations. This approximation looks like the inverse of a linear operator as mentioned in Ref. [1].

In accordance with the literature, many researchers have contributed greatly in investigating the area of studying pseudo-differential operators [2–4]. These operators have been studied on several topological spaces depending on their importance in the theory of PDE. Also, it is quite important to mention that the study of pseudo-differential operators on the Euclidean space $\mathbb{R}^n$ characterized them as standard pseudo-differential operators, and they are presented in many books [5, 6].

The properties of pseudo-differential operators on some topological spaces such as Sobolev space, matrix group SU(2), homogeneous space, and compact Lie groups are investigated in Ref. [7]. A proposed study on nuclear pseudo-differential operators in Borel measures is investigated in Refs. [8, 9]. Another approach is proposed for showing the boundedness of pseudo-differential operators.

A study of pseudo-differential operators is investigated in $\mathbb{Z}$ and $\mathbb{S}^1$ in Refs. [1, 10]. In this, a necessary and sufficient condition imposed on the measurable function $\sigma : \mathbb{Z} \times \mathbb{S}^1 \to \mathbb{C}$ to guarantee that the corresponding pseudo-differential operators are Hilbert-Schmidt operators, L^p-boundedness and L^p-compactness. Finding good conditions on the symbol to guarantee compactness and boundedness of this kind of operators is also part of the interesting problems.

The theory of pseudo-differential operators forms a class denoted as Ψ^m and is considered as an extension of the class of linear differential operators where m is the order of the operators. The theory of pseudo-differential operators is studied in mathematical analysis [11] and is very important in solving elliptic differential equations. Around 1880, this theory was of relevant importance in the theory of nonlinear partial differential operators [4].

In order to motivate the definition of pseudo-differential operators, the theory of Fourier transform and that of symbol classes are fundamental tools to build pseudo-differential operators and their properties.

As a simple illustrating example, we give a structure in constructing pseudo-differential operators on the Euclidean space.

We consider $\mathcal{S} \equiv \mathcal{S}(\mathbb{R}^n)$ to be the set of smooth functions $\mathbb{R}^n \to \mathbb{C}$ that rapidly decay at infinity. For $\varphi \in \mathcal{S}$, its Fourier transform is given by

$$\hat{\varphi}(\xi) = \int_{\mathbb{R}^n} e^{-2\pi i x \cdot \xi} \varphi(x) \mathrm{d}x, \tag{1}$$

where $x, \xi \in \mathbb{R}^n$ and $x \cdot \xi = \sum_{i=1}^n x_i \xi_i$ (where the $\cdot$ stands for the scalar product). The equation above represents the Fourier inversion formula of the function φ given without proof.

That is, if $\hat{\varphi}$ is the Fourier transform of the function $\varphi \in \mathcal{S}$, the Fourier inversion formula is given by

$$\varphi(x) = \int_{\mathbb{R}^n} e^{2\pi i x \xi} \hat{\varphi}(\xi) \mathrm{d}\xi, \tag{2}$$

for all $x \in \mathbb{R}^n$.

By the Fourier inversion formula, we can define the standard pseudo-differential operators (in $\mathbb{R}^n$). To start with, we consider the general linear partial differential operator P of order m on $\mathbb{R}^n$ which is given by

$$P(x, D) = \sum_{|\alpha| \leq m} a_\alpha(x) D^\alpha, \tag{3}$$

where $a_\alpha(x)$ are functions defined on $\mathbb{R}^n$ and α is a multi-index whose length is $|\alpha| = \alpha_1 + \ldots + \alpha_n$. We could also use $\frac{\partial^\alpha}{\partial}$ instead of D^α. When we replace D^α by the monomial of the form $\xi^\alpha \in \mathbb{R}$, we obtain the so-called symbol (see Ref. [5]) defined by

$$p(x, \xi) = \sum_{|\alpha| \leq m} a_\alpha(x) \xi^\alpha \tag{4}$$

of the operator P defined in (3).

Once we have defined the Fourier transform and its inversion formula, we can now construct pseudo-differential operators.

We take f to be another Schwartz function. By Eqs. (3) and (2) and (4), we obtain the following equations.

$$P(x, D) f(x) = \sum_{|\alpha| \leq m} a_\alpha(x) (D^\alpha f)(x) \tag{5}$$

$$= \sum_{|\alpha|\leq m} a_\alpha(x)\left(\int_{\mathbb{R}^n} e^{2\pi i x\cdot\xi}(\widehat{D^\alpha f})(\xi)\mathrm{d}\xi\right) \tag{6}$$

$$= \sum_{|\alpha|\leq m} a_\alpha(x)\left(\int_{\mathbb{R}^n} e^{2\pi i x\cdot\xi}\xi^\alpha\hat{f}(\xi)\mathrm{d}\xi\right) \tag{7}$$

$$= \int_{\mathbb{R}^n} e^{2\pi i x\cdot\xi}\left(\sum_{|\alpha|\leq m} a_\alpha(x)\xi^\alpha\right)\hat{f}(\xi)d\xi \tag{8}$$

$$= \int_{\mathbb{R}^n} e^{2\pi i x\cdot\xi}p(x,\xi)\hat{f}(\xi)d\xi. \tag{9}$$

By this, we may see that the last line is an illustration of a differential operator expressed in terms of $p(x, \xi)$ which are called symbols. This also shows that it is possible to construct an operator more general than any linear partial differential operators. Furthermore, when we replace $p(x, \xi)$ with the general symbol $\sigma(x, \xi)$, we have

$$P(x, D)f(x) = \int_{\mathbb{R}^n} e^{2\pi i x\cdot\xi}\sigma(x,\xi)\hat{f}(\xi)\mathrm{d}\xi. \tag{10}$$

The operator P obtained so is generally called a pseudo-differential operator corresponding to the function $\sigma(x, \xi)$ called the symbol.

Pseudo-different operators qualify as a generalization of linear differential operators that depend only on their symbols. These symbols are or are not necessarily polynomial functions. The set of symbols of pseudo-differential operators on $\mathbb{R}^n$ is denoted by $S^m(\mathbb{R}^n \times \mathbb{R}^n)$ where m is the order of the operator. The definition of this set is provided in the Preliminaries section of this chapter.

We outline this chapter in four sections. In Section 1, we provide a preliminary description of some definitions and relevant notions that we recall in order to achieve the main goal. In Section 2, we present essential tools on Fourier transforms on $\mathbb{Z}^N$ that are useful for the more details of the main part of the chapter which is mentioned in Section 3. Section 4 is the conclusion.

2. Preliminaries

Definition 2.1 (L^p space). Let $1 \leq p < \infty$ and (S, Σ, μ) be a measurable space. The L^p space may be defined as set of all measurable functions $f : S \to \mathbb{C}$ such that

$$\|f\|_p \equiv \left(\int_S |f|^p d\mu\right)^{\frac{1}{p}} < \infty. \tag{11}$$

The following operation holds for the set of such functions:

- $(f + g)(u) = f(u) + g(u)$
- $(\lambda f)(u) = \lambda f(u)$

Definition 2.2 (Banach space). A normed vector space that is complete with respect to the metric induced by the norm is called a Banach space.

Definition 2.3 (Inner product and Hilbert space). Let $\mathcal{H}$ be a vector space over the field $\mathbb{C}$. A mapping $(\langle u, v\rangle \mapsto \langle u, v\rangle) : \mathcal{H} \times \mathcal{H} \to \mathbb{C}$ is an inner product if.

- $\langle u+v,w\rangle=\langle u,w\rangle+\langle v,w\rangle$
- $\langle \lambda u,v\rangle=\lambda\langle u,v\rangle$
- $\langle u,v\rangle=\overline{\langle v,u\rangle}$
- $\langle u,u\rangle\geq 0$
- $\langle u,u\rangle=0\rightarrow u=0$

for all $u,v\in\mathcal{H}$ and $\lambda\in\mathbb{C}$. Then $\mathcal{H}$ endowed with the inner product is the *inner product space*. An inner product defines the canonical norm

$$\|u\|:=\langle u,u\rangle^{\frac{1}{2}} \tag{12}$$

If $\mathcal{H}$ is a Banach space with respect to the canonical form, then it is called a *Hilbert space* (or a complete inner product).

Definition 2.4 (Operator). A mapping that acts on elements of a particular space to generate other elements in this same space is referred to as an *operator*.

Definition 2.5 (Linear operators). Let V and W be vector spaces over a field K. An operator $Q:V\rightarrow W$ is a linear operator if $\forall x,y\in V$ and $\alpha,\beta\in K$ we have

$$Q(\alpha x+\beta y)=\alpha Qx+\beta Qy \tag{13}$$

Linear operators are necessary since they preserve the properties of the space it acts on (say continuity). It is the most common kind of operator. Examples of linear operator are integral operators, differential operators, and Fourier transforms among others.

Definition 2.6 (Bounded operators). A linear operator from V to W is bounded if there exists $M>0$ such that $\forall x\in V$

$$\|Qx\|_W\leq M\|x\|_V \tag{14}$$

where V and W are vector spaces over the same ordered field (say $\mathbb{R}$) and are associated with norms.

Definition 2.7 (Differential operator). In mathematics, a differential operator P of order r on the Euclidean space is a polynomial in the derivatives $\partial_x=(\partial_{x1},\partial_{x2},\ldots,\partial_{xn})$ which is expressed as

$$P=\sum_{|\alpha|\leq r}c_\alpha(x)\partial_x^\alpha=\sum_{|\alpha|\leq r}c_\alpha(x)D^\alpha, \tag{15}$$

where c_α is a constant depending on x, to its *symbol*

$$a(x,\xi)=\sum_{|\alpha|\leq r}c_\alpha(x)(i\xi)^\alpha. \tag{16}$$

Definition 2.8 (Symbol). The symbol of a linear differential operator is a polynomial in the *phase variable* $\xi\in\mathbb{R}^n$ whose constants depend on the *space variable* x. It partially controls the qualitative behavior of solutions of a partial differential eq. A symbol associates to a differential operator by replacing each partial derivative by a new variable (say *xi*). It is mainly applied in Fourier analysis. The notion of a pseudo-differential operator is motivated by the symbol in connection to the Fourier transforms.

Definition 2.9 (Symbol classes S^m). A function $\lambda \in S^m$ is referred to as a symbol if for $m \in (-\infty, \infty)$, S^m is the set of all the functions $\lambda \in C^\infty(\mathbb{Z}^n \times \mathbb{Z}^n)$ so that for all multi-indices α, ψ, $\exists$ a positive constant $C_{\alpha,\psi}$ whereby

$$|D_x^\psi D_\xi^\alpha \lambda(x,\xi)| \le C_{\alpha,\psi}(1+|\xi|)^{m-|\alpha|} \qquad x,\xi \in \mathbb{Z}^n \tag{17}$$

Definition 2.10 (Principal symbol). A principal symbol (a) is the highest order terms of the symbol, that is

$$a(x,\xi) = \sum_{|\alpha|=r} c_\alpha(x)(i\xi)^\alpha. \tag{18}$$

Linear partial differential equations whose principal symbols is nowhere zero can be called elliptic partial differential equations. Also for hyperbolic and parabolic partial differential equations, zeros of the principal symbol is associated with the characteristics of the partial differential equation. In particular, the symbol is mostly fundamental for the solution of these equations, and an essential computational devices are used to study their singularities.

Definition 2.11 (Space $L^p(\Omega)$, $1 \le p < \infty$). Let Ω be an open set of $\mathbb{R}^n$, p be a positive real number and let (S,μ) be a measurable space. $L^p(\Omega)$ is the set of measurable functions whose absolute value raised to the $p-$th power has a finite Lebesgue integral given by

$$\|f\|_{L^p(\Omega)} := \left(\int_\Omega |f|^p \mathrm{d}\mu\right)^{1/p} < \infty. \tag{19}$$

Remark 2.12.

- If $p = 1$, we have the following

$$\|f\|_{L^1(\Omega)} = \int_\Omega |f| \mathrm{d}\mu, \tag{20}$$

where $L^1(\Omega)$ is the set of absolutely integrable functions on Ω.

- In case $p = 2$ and $\Omega = \mathbb{R}^n$, the space $L^2(\mathbb{R}^n)$ is the usual Hilbert space which consists of the square-integrable measurable functions on $\mathbb{R}^n$. In the following are some properties of the Hilbert space.

Proposition 2.13 (Elementary properties of $L^2(\mathbb{R}^n)$).

- The space of functions $L^2(\mathbb{R}^n)$ is a linear space.
- The norm defined on this space satisfies Minkwoski's inequality.
- The topology on $L^2(\mathbb{R}^n)$ is given by an homogeneous norm of degree 1.
- The linear space $L^2(\mathbb{R}^n)$ is complete with respect to the norm

$$\|f\|_{L^2(\mathbb{R}^n)} = \left(\int_\Omega |f(x)|^2 \mathrm{d}x\right)^{1/2}. \tag{21}$$

That is if $\{f_k\}_{k=1}^n \subseteq L^2(\mathbb{R}^n)$ is a Cauchy sequence of functions, then that sequence converges, that is, there exists a limit function $f(x) \in L^2(\mathbb{R}^n)$.

(e) The norm on $L^2(\mathbb{R}^n)$ might also be given by an inner product,

$$\langle f,g\rangle = \int_{\mathbb{R}^n} f(x)\overline{g(x)}\mathrm{d}x, \quad \langle f,f\rangle = \|f\|^2_{L^2(\mathbb{R}^n)}. \tag{22}$$

The inner product satisfies the Cauchy-Schwartz's inequality (see (3.4)).

(f) The "Dual Space" or space of bounded functional of $L^2(\mathbb{R}^n)$ is $L^2(\mathbb{R}^n)$ itself.

Definition 2.14 (Measurable function). Let (X,A) and (Y,B) be two measurable spaces. The function $f : X \to Y$ is said to be measurable if for all $B \in X, f^{-1}(B) \in Y$.

Measurability of a function depends only on the σ-algebras.

To show that a function is measurable, it is enough to check the measurability of the inverse images of sets that generate the σ-algebra.

3. Some theorems and useful inequalities on $L^p(\Omega)$

Theorem 3.1 (Hölder's Inequality). *Let $1 \leq p \leq \infty$, where $\frac{1}{p}+\frac{1}{q}=1$. Let $f \in L^p(\Omega)$ and $g \in L^q(\Omega)$. Then $fg \in L^1(\Omega)$*

$$\|fg\|_{L^1(\Omega)} \leq \|f\|_{L^p(\Omega)}\|g\|_{L^q(\Omega)}. \tag{23}$$

The above inequality is the generalization of Cauchy-Schwartz inequality.

Proof: If $p=1$ or $p=\infty$ then the inequality is trivial. Suppose $1 \leq p \leq \infty$ and let

$$a = \int_\Omega |f|^p d\mu, \qquad b = \int_\Omega |g|^q d\mu, \tag{24}$$

setting

$$u = \frac{|f|}{a^{\frac{1}{p}}}, \qquad v = \frac{|g|}{b^{\frac{1}{q}}} \tag{25}$$

By Young's inequality

$$uv = \left(\frac{|f|^p}{a}\right)^{\frac{1}{p}}\left(\frac{|g|^q}{b}\right)^{\frac{1}{q}} \leq \frac{1}{p}\left(\frac{|f|^p}{a}\right) + \frac{1}{q}\left(\frac{|g|^q}{b}\right) \tag{26}$$

$$\frac{1}{a^{\frac{1}{p}}b^{\frac{1}{q}}}|f||g| \leq \frac{1}{pa}|f|^p + \frac{1}{qb}|g|^q \tag{27}$$

Integrating both sides

$$\frac{1}{a^{\frac{1}{p}}b^{\frac{1}{q}}}\int_\Omega |f||g|d\mu \leq \frac{1}{pa}\int_\Omega |f|^p d\mu + \frac{1}{qb}\int_\Omega |g|^q d\mu \tag{28}$$

$$\frac{1}{\|f\|_{L^p(\Omega)}\|g\|_{L^q(\Omega)}}\int_\Omega |f||g|d\mu \leq \frac{1}{pa}a + \frac{1}{qb}b \tag{29}$$

$$= \frac{1}{p} + \frac{1}{q} \tag{30}$$

$$= 1, \tag{31}$$

which implies that

$$\int_\Omega |f||g|d\mu \leq \|f\|_{L^p(\Omega)}\|g\|_{L^q(\Omega)}. \tag{32}$$

Therefore the desired inequality $\|fg\|_{L^1(\Omega)} \leq \|f\|_{L^p(\Omega)}\|g\|_{L^q(\Omega)}$.

Theorem 3.2 (Minkowki's Inequality). *Let* $1 \leq p \leq \infty$ *and* $f, g \in L^p(\Omega)$. *Then,*

$$\|f+g\|_{L^p(\Omega)} \leq \|f\|_{L^p(\Omega)} + \|g\|_{L^p(\Omega)} \tag{33}$$

Proof: If $p = 1$ or $p = \infty$ then from the triangle inequality for complex numbers, it is trivial. Suppose $1 \leq p \leq \infty$, thus we have

$$\|f+g\|^p_{L^p(\Omega)} = \int_\Omega |f+g|^p du \tag{34}$$

$$= \int_\Omega |f+g|^{p-1}|f+g|du \tag{35}$$

$$\leq \int_\Omega |f+g|^{p-1}(|f| + |g|)du \tag{36}$$

$$= \int_\Omega |f+g|^{p-1}|f|du + \int_\Omega |f+g|^{p-1}|g|du \tag{37}$$

$$\leq \left(\int_\Omega |f+g|^{p-1}du\right)\left(\int_\Omega |f|du\right) + \left(\int_\Omega |f+g|^{p-1}du\right)\left(\int_\Omega |g|du\right) \tag{38}$$

$$= \left[\int_\Omega |f+g|^{p-1}du\right]\left[\int_\Omega |f|du) + \int_\Omega |g|du\right]. \tag{39}$$

Using Hölder's inequality where $p = p, q = \frac{p}{p-1}$ then it implies

$$\|f+g\|^p_{L^p(\Omega)} \leq \int_\Omega [|f+g|^p du]^{\frac{p-1}{p}}\left[\left(\int_\Omega |f|^p du\right)^{\frac{1}{p}} + \left(\int_\Omega |g|^p du\right)^{\frac{1}{p}}\right] \tag{40}$$

$$= \|f+g\|^{p-1}_{L^p(\Omega)}\left(\|f\|_{L^p(\Omega)} + \|g\|_{L^p(\Omega)}\right), \tag{41}$$

This means that

$$\|f+g\|_{L^p(\Omega)} \leq \|f\|_{L^p(\Omega)} + \|g\|_{L^p(\Omega)}. \tag{42}$$

3.1 Young's inequality

Let p, q be two conjugate real numbers, that is, $\frac{1}{p} + \frac{1}{q} = 1$ for all $p, q \in (1, \infty)$. Then,

$$ab \leq \frac{1}{p}a^p + \frac{1}{q}b^q, \quad \text{for all,} \quad a, b > 0. \tag{43}$$

Proof: It is straightforward to show that for $p = q = 2$, the inequality

$$ab \leq \frac{1}{2}a^2 + \frac{1}{2}b^2, \quad \text{for all } a, b > 0, \tag{44}$$

holds. However, considering the convexity of the logarithm function, which says that its graph lies above the line segment connecting two points on the graph,

$$t\log x + (1-t)\log y \leq \log(tx + (1-t)y), \quad \text{for } x, y > 0 \text{ and } \quad 0 \leq t \leq 1. \tag{45}$$

Then, we have

$$\frac{1}{p}\log a^p + \left(1 - \frac{1}{p}\right)\log b^q \leq \log\left(\frac{1}{p}a^p + \left(1 - \frac{1}{p}\right)b^q\right) \tag{46}$$

For $t = \frac{1}{p}$; $x = a^p$ and $y = b^q$, we obtain

$$\frac{1}{p}\log a^p + \frac{1}{q}\log b^q \leq \log\left(\frac{1}{p}a^p + \frac{1}{q}b^q\right) \tag{47}$$

$$\log a + \log b \leq \log\left(\frac{1}{p}a^p + \frac{1}{p}b^q\right). \tag{48}$$

Hence, the proof is complete. The consequence of this inequality gives for $f \in L^p(\Omega)$ and $g \in L^q(\Omega), fg \in L^1(\Omega)$ and

$$\|fg\|_{L^1(\Omega)} \leq \frac{1}{p}\|f\|^p_{L^p(\Omega)} + \frac{1}{q}\|g\|^q_{L^p(\Omega)}. \tag{49}$$

3.2 Cauchy-Schwartz's inequality

The Cauchy-Schwartz's inequality is given by

$$\|fg\|^2_{L^1} \leq \|f\|^2_{L^2(\Omega)}\|g\|^2_{L^2(\Omega)} \tag{50}$$

It is also considered as a particular case of the Hölder's inequality when $p = q = 2$.

Remark 3.5. A Banach space whose norm is given by an inner product is a Hilbert space. Our favorite Hilbert space $L^2(\mathbb{R}^n)$ has its inner product. This is one reason why $L^2(\mathbb{R}^n)$ is a natural setting for the Fourier transform. This reason leads us to the next chapter on Fourier transforms that are very useful for the theory of ΨDOs.

4. Some useful information about the Fourier transform

In this section, we introduce the definition of Fourier transform on $L^1(\mathbb{R}^N)$ and $L^2(\mathbb{Z}^N)$ and review some basic elements. Also, we emphasize on certain properties of the Fourier transform that would be used as part of the main focus.

4.1 Definitions

Definition 4.1 (L^p space). Let $1 \leq p < \infty$. A function $f : \Omega \to \mathbb{C}$ is said to be $L^p(\Omega)$ if it is measurable and its norm

$$\|f\|_{L^p(\Omega)} := \left(\int_\Omega |f(u)|^p du \right)^{\frac{1}{p}} < \infty. \tag{51}$$

When $p = \infty$, then f is said to be in $L^\infty(\Omega)$ it is essentially bounded and measurable.

$$\|f\|_{L^\infty(\Omega)} := \text{esssup}_{u \in \Omega} |f(u)| < \infty. \tag{52}$$

where $\text{esssup}_{u \in \Omega} |f(u)|$ is the smallest M such that $|f(u)| \leq M$ for almost all $u \in \Omega$.

Definition 4.2 (Fourier transform on $L^1(\mathbb{R}^N)$). Let $f \in L^1(\mathbb{R}^N)$, then its Fourier transform $\mathscr{F}_{\mathbb{Z}^N} b$ is a function on $\mathbb{R}^N$ which is given by

$$(\mathscr{F}_{\mathbb{R}^N} f)(\varepsilon) = \int_{\mathbb{R}^N} f(x) e^{-2\pi i x \cdot \varepsilon} dx. \tag{53}$$

Definition 4.3 (Fourier transform on $L^2(\mathbb{Z}^N)$). Let $b(n) \in L^2(\mathbb{Z}^N)$, the Fourier transform $\mathcal{F}_{\mathbb{Z}^N} b$ is a function on the torus $\mathbb{T}^N$ which is given by

$$(\mathcal{F}_{\mathbb{Z}^N} b)(\phi) = \sum_{n \in \mathbb{Z}^N} b(n) e^{-in \cdot \phi}, \qquad \phi \in \mathbb{T}^N \tag{54}$$

where $n = (n_1, n_2, \cdots, n_N) \in \mathbb{Z}^2$, $\phi = (\phi_1, \phi_2, \cdots, \phi_N) \in \mathbb{T}^2$ and $n \cdot \phi = \sum_{j=1}^{N} n_j \phi_j$. The Fourier transform can also be represented as

$$(\mathcal{F}_{\mathbb{Z}^N} b)(\phi) = \frac{1}{2\pi} \int_{\mathbb{Z}^N} e^{-in \cdot \phi} b(n) dn, \tag{55}$$

and the *inverse Fourier transform* is given by

$$b(n) = \frac{1}{2\pi} \int_{\mathbb{T}^N} e^{in \cdot \phi} (\mathcal{F}_{\mathbb{Z}^2} b)(\phi) d\phi, \qquad n \in \mathbb{Z}^N. \tag{56}$$

Remark 4.4. The expression of the Fourier transform is written in various ways in other literature, that is the formula may have the constant $\frac{1}{2\pi}, \frac{1}{\sqrt{2\pi}}$ or the exponential may be $e^{-i\pi n \phi}, e^{-2i\pi n \phi}$ among others. If the exponential is in the form $e^{-2i\pi n \phi}$, then there will be no constant coefficient.

4.1.1 Some properties of Fourier transform

Definition 4.5 (Multi-indices). In $\mathbb{R}^N$, we mainly use a multi-index notation. For multi-indices $\alpha = (\alpha_1, \alpha_2, \cdot, \alpha_N)$ and $\delta = (\delta_1, \delta_2, \cdots, \delta_N)$ where the integers $\alpha_j, \delta_j \geq 0$, we define

$$\partial^{\alpha}\psi(x) = \frac{\partial^{\alpha_1}}{\partial^{\alpha_1}_{x_1}} \cdots \frac{\partial^{\alpha_N}}{\partial^{\alpha_N}_{x_N}}\psi(x) \tag{57}$$

and $x^{\delta} = x_1^{\delta_1} \cdots x_N^{\delta_N}$. We write $\alpha \leq \delta$ for multi-indices α and δ which means that $\alpha_j \leq \delta_j$ for all $j \in \{1, \cdots, N\}$. Also the length of the multi-index α is denoted by
$|\alpha| \ = \alpha_1 + \alpha_2 + \cdot + \alpha_N$.

Definition 4.6 (Schwartz space $\mathcal{S}(\mathbb{Z}^N)$). Let $\mathcal{S}(\mathbb{Z}^N)$ represent Schwartz space which is space of rapidly decreasing functions $\mathbb{Z}^N \to \mathbb{C}$. Then $\psi \in \mathcal{S}(\mathbb{Z}^N)$, if for any $K < \infty$, $\exists$ a constant $C_{\psi,K}$ such that the relation

$$|\psi(\xi)| \ \leq C_{\psi,K}\langle\xi\rangle^{-K} \tag{58}$$

is true for all $\xi \in \mathbb{Z}^N$, where $\langle\xi\rangle = (1+|\xi|)^{\frac{1}{2}}$.

Theorem 4.7 (Fourier inversion formula [7]**).** *The Fourier transform $\mathcal{F} : \varphi \mapsto \hat{\varphi}$ is an isomorphism of $\mathcal{S}(\mathbb{R}^N)$ into $\mathcal{S}(\mathbb{R}^N)$ whose inverse is given by*

$$\varphi(x) = \int_{\mathbb{R}^N} e^{2\pi i x\cdot\xi}\hat{\varphi}(\xi)d\xi. \tag{59}$$

The formula 59 is referred to as the Fourier inversion formula. Also the inverse Fourier transform is denoted by

$$\left(\mathcal{F}^{-1}_{\mathbb{R}^N}f\right)(x) \equiv \left(\mathcal{F}^{-1}f\right)(x) := \int_{\mathbb{R}^N} e^{2\pi i x\cdot\xi}f(\xi)d\xi. \tag{60}$$

Thus, we say that

$$\mathcal{F}\circ\mathcal{F}^{-1} = \mathcal{F}^{-1}\circ\mathcal{F} = identity \tag{61}$$

on $\mathcal{S}(\mathbb{R}^N)$.

Theorem 4.8 (Fubini's Theorem). *Suppose A and B are complete measure spaces. Suppose $f(u,v)$ is $A \times B$ measurable. If*

$$\int_{A\times B} |f(u,v)| d(u,v) < \infty \tag{62}$$

with the integral is taken with respect to a product measure on the space over $A \times B$, then

$$\int_A \left(\int_B f(u,v)du\right)dv = \int_B \left(\int_A f(u,v)dv\right)du = \int_{A\times B} f(u,v)d(u,v) \tag{63}$$

Corollary 4.9. If $f(u,v) = g(u)h(v)$ for some functions g and h then

$$\int_A g(u)du \int_B h(v)dv = \int_{A\times B} f(u,v)d(u,v), \tag{64}$$

where the integral on the right side is with respect to a product measure.

Lemma 4.10 (Multiplication formula for the Fourier transform). Let $f, g \in L^1(\mathbb{R}^N)$. Then

$$\int_{\mathbb{R}^N} \hat{f} g du = \int_{\mathbb{R}^N} f \hat{g} dv. \tag{65}$$

Proof: By definition we have

$$\int_{\mathbb{R}^N} \hat{f} g du = \int_{\mathbb{R}^N} \left[\int_{\mathbb{R}^N} e^{-2\pi i u \cdot v} f(v) dv \right] g(u) du \tag{66}$$

Applying Fubini's theorem (4.8)

$$\int_{\mathbb{R}^N} \hat{f} g du = \int_{\mathbb{R}^N} \left[\int_{\mathbb{R}^N} e^{-2\pi i u \cdot v} g(u) du \right] f(v) dv \tag{67}$$

$$= \int_{\mathbb{R}^N} f \hat{g} dv \tag{68}$$

Hence the proof.

Definition 4.11 (Convolutions). Let the functions $f, g \in L^1(\mathbb{R}^N)$, their convolution is defined by

$$(f^* g)(u) := \int_{\mathbb{R}^N} f(u - v) g(v) dv \tag{69}$$

The convolution $f^* g \in L^1(\mathbb{R}^N)$ with the norm inequality satisfies

$$\|f^* g\|_{L^1(\mathbb{R}^N)} \leq \|f\|_{L^1(\mathbb{R}^N)} \|g\|_{L^1(\mathbb{R}^N)} Je \tag{70}$$

Remark 4.12. In particular, the convolution can be defined rigorously by first defining (69) for $f, g \in \mathcal{S}(\mathbb{R}^N)$ and then extending it to a mapping $^* : L^1(\mathbb{R}^N) \times L^1(\mathbb{R}^N) \to L^1(\mathbb{R}^N)$ by (70) this ensures the convergence of the integral in (69).

The following properties relate convolutions with Fourier transforms.

Theorem 4.13. *Let* $\varphi, \psi \in \mathcal{S}(\mathbb{R}^N)$. *Then we have*

(i) $\int_{\mathbb{R}^N} \varphi \overline{\psi} dx = \int_{\mathbb{R}^N} \hat{\varphi} \overline{\hat{\psi}} d\varepsilon.$

(ii) $\varphi \hat{*} \psi(\varepsilon) = \hat{\varphi}(\varepsilon) \hat{\psi}(\varepsilon).$

(iii) $\widehat{\varphi\psi}(\varepsilon) = (\hat{\varphi}^* \hat{\psi})(\varepsilon).$

Proof: (i) Given that

$$\mathcal{F}(\psi)(\varepsilon) = \hat{\psi}(\varepsilon) = \int_{\mathbb{R}^N} e^{-2\pi i x \cdot \varepsilon} \psi(x) dx \tag{71}$$

we denote

$$Y(\varepsilon) = \overline{\hat{\psi}}(\varepsilon) = \int_{\mathbb{R}^N} e^{2\pi i x\cdot\varepsilon}\overline{\psi}(x)dx = \mathcal{F}^{-1}(\overline{\psi})(\varepsilon), \tag{72}$$

thus $\hat{Y} = \overline{\psi}$. This implies that

$$\int_{\mathbb{R}^N} \varphi\overline{\psi}dx = \int_{\mathbb{R}^N} \varphi\hat{Y}dx \tag{73}$$

$$= \int_{\mathbb{R}^N} \hat{\varphi}Ydx \tag{74}$$

$$= \int_{\mathbb{R}^N} \hat{\varphi}\overline{\hat{\psi}}d\varepsilon \tag{75}$$

where we used the multiplication formula for the Fourier transform in Lemma 4.1.5.

(ii) It is simple to compute the second property

$$\hat{\varphi} * \psi(\varepsilon) = \int_{\mathbb{R}^N} e^{-2\pi i x\cdot\varepsilon}(\varphi * \psi)(x)dx \tag{76}$$

$$= \int_{\mathbb{R}^N}\int_{\mathbb{R}^N} e^{-2\pi i x\cdot\varepsilon}\varphi(x-y)\psi(y)dydx \tag{77}$$

$$= \int_{\mathbb{R}^N}\int_{\mathbb{R}^N} e^{-2\pi i(x-y)\cdot\varepsilon}\varphi(x-y)e^{-2\pi i y\cdot\varepsilon}\psi(y)dydx \tag{78}$$

$$= \int_{\mathbb{R}^N}\int_{\mathbb{R}^N} e^{-2\pi i z\cdot\varepsilon}\varphi(z)e^{-2\pi i y\cdot\varepsilon}\psi(y)dydz \tag{79}$$

$$= \hat{\varphi}(\varepsilon)\hat{\psi}(\varepsilon) \tag{80}$$

that is by substituting $z = x - y$.

(iii)

$$\hat{\varphi\psi}(\varepsilon) = \int_{\mathbb{R}^N} e^{-2\pi i x\cdot\varepsilon}\varphi(x)\psi(x)dx \tag{81}$$

$$= \left(\int_{\mathbb{R}^N} e^{-2\pi i x\cdot\varepsilon}\varphi(x)dx\right)\left(\int_{\mathbb{R}^N} e^{2\pi i y\cdot\varepsilon}\hat{\psi}(y)dy\right) \quad \text{by substituting } \psi(x) \tag{82}$$

$$= \int_{\mathbb{R}^N}\underbrace{\int_{\mathbb{R}^N} e^{-2\pi i(x-y)\cdot\varepsilon}\varphi(x)dx}_{\hat{\varphi}(x-y)}\hat{\psi}(y)dy \tag{83}$$

$$= \int_{\mathbb{R}^N} \hat{\varphi}(x-y)\hat{\psi}(y)dy \tag{84}$$

$$= (\hat{\varphi} * \hat{\psi})(\varepsilon) \tag{85}$$

This completes the proof.

Theorem 4.14 (Plancherel's and Parserval's formula). *Let $u \in L^2(\mathbb{R}^N)$. Then $\hat{u} \in L^2(\mathbb{R}^N)$ and we have*

$$\|\hat{u}\|_{L^2(\mathbb{R}^N)} = \|u\|_{L^2(\mathbb{R}^N)} \qquad \left(Plancherel's\ identity\right) \tag{86}$$

Also for all $u, v \in L^2(\mathbb{R}^N)$,

$$\int_{\mathbb{R}^N} u\bar{v}dx = \int_{\mathbb{R}^N} \hat{u}\bar{\hat{v}}d\xi \qquad \left(Parseval's\ identity\right) \tag{87}$$

5. Pseudo-differential operators on $\mathbb{Z}^N$

A lot of study has been done on pseudo-differential operators on $\mathbb{R}^N$ and $\mathbb{T}^N$. Also, various theorems and proofs of the conditions imposed on the symbol $\sigma : \mathbb{Z} \times \mathbb{S}^1 \to \mathbb{C}$ to guarantee that the corresponding Pseudo-differential operator $T_\sigma : L^2(\mathbb{Z}) \to L^2(\mathbb{Z})$ are Hilbert-Schmidt, bounded or compact is in Ref. [1]. In this chapter, we introduce the composition of two Pseudo-differential on $\mathbb{Z}^N$, its amplitude and kernel. We also explore and extend the theorems and proofs in Ref. [10] to the integer lattice ($\mathbb{Z}^N$).

6. Notions and definitions

We begin informally by saying if T a translation invariant linear operator on $\mathbb{Z}^N$, then

$$T_\sigma\left(e^{in\cdot\phi}\right) = \sigma(n,\phi)e^{in\cdot\phi} \qquad \text{for all} \qquad \phi \in \mathbb{T}^N. \tag{88}$$

Also, if T acts on functions with a different variable (say m), we set

$$f(n,\phi) = T_\sigma\left(e^{im\cdot\phi}\right)(n) = (T_\sigma c)(n), \qquad \text{with} \quad c(n) = e^{in\cdot\phi}. \tag{89}$$

From (88), we have

$$\sigma(n,\phi) = e^{-in\cdot\phi}T_\sigma\left(e^{in\cdot\phi}\right) \tag{90}$$

Now, we consider when T is applied to the Fourier inversion formula. Let b be a function of $L^1(\mathbb{Z}^N)$ and the Fourier transform $\mathcal{F}_{\mathbb{Z}^N}b$ of b is a function on the torus $\mathbb{S}^1 \times \mathbb{S}^1 \times \cdots \mathbb{S}^1 = \mathbb{T}^N$ defined by

$$(\mathcal{F}_{\mathbb{Z}^N}b)(\phi) = \sum_{n\in\mathbb{Z}^N} b(n)e^{-in\cdot\phi}, \qquad \phi \in \mathbb{T}^N \tag{91}$$

with $n = (n_1, n_2, \cdots, n_N) \in \mathbb{Z}^N$, $\phi = (\phi_1, \phi_2, \cdots, \phi_N) \in \mathbb{T}^N$, and $n \cdot \phi = \sum_{j=1}^{N} n_j\phi_j$.

Where $\mathcal{F}_{\mathbb{Z}^N}$ extends to $L^2(\mathbb{Z}^N)$ with $\mathcal{F}_{\mathbb{Z}^2}b \in L^2(\mathbb{T}^N)$, and by Plancherel's formula for Fourier series we have

$$\sum_{n\in\mathbb{Z}^N} |b(n)|^2 = \frac{1}{2\pi}\int_{\mathbb{T}^N} \left|(\mathcal{F}_{\mathbb{Z}^N}b)(\phi)\right|^2 d\phi. \tag{92}$$

Using Eq. (91), the inversion formula for Fourier series gives

$$b(n) = \frac{1}{2\pi}\int_{\mathbb{T}^N} e^{in\cdot\phi}(\mathcal{F}_{\mathbb{Z}^N} b)(\phi)d\phi, \qquad n\in\mathbb{Z}^N. \tag{93}$$

Definition 6.1 (Pseudo-differential operator on $\mathbb{Z}^N$). Let consider a measurable function σ such that $\sigma : \mathbb{Z}^N \times \mathbb{T}^N \to \mathbb{C}$ and for $b\in\mathcal{S}(\mathbb{Z}^N)$, the sequence $T_\sigma b$ is given by

$$(T_\sigma b)(n) = \frac{1}{2\pi}\int_{\mathbb{T}^N} e^{in\cdot\phi}\sigma(n,\phi)(\mathcal{F}_{\mathbb{Z}^N} b)(\phi)d\phi, \qquad n\in\mathbb{Z}^N. \tag{94}$$

Definition 6.2 (Symbol classes $S^m(\mathbb{Z}^N \times \mathbb{T}^N)$). A function $\sigma\in S^m(\mathbb{Z}^N \times \mathbb{T}^N)$ if for $m\in(-\infty,\infty)$, $\sigma = \sigma(n,\phi)$ is smooth on $\mathbb{Z}^N \times \mathbb{T}^N$ so that for all multi-indices α,ψ, $\exists$ a positive constant $C_{\alpha,\psi}$ whereby

$$|\partial_n^\psi \partial_\phi^\alpha \sigma(n,\phi)| \le C_{\alpha,\psi}(1+|\phi|)^{m-|\alpha|} \qquad n\in\mathbb{Z}^n, \ \phi\in\mathbb{T}^n \tag{95}$$

The operator defined by Eq. (94) is called the pseudo-differential operator on $\mathbb{Z}^N$ corresponding to the symbol σ. This pseudo-differential operator is a map from $\mathcal{S}(\mathbb{Z}^N)$ to $\mathcal{S}(\mathbb{Z}^N)$. The formula (94) permits us to simplify some properties of the operator T_σ to properties of multiplication by the symbol $\sigma(n,\phi)$ associated with T. For instance, the continuity of T_σ on L^2 would be simplified to the boundedness of $\sigma(n,\phi)$, composition of two operators $T_{\sigma_1}\circ T_{\sigma_2}$ would reduce to the multiplication of their respective symbols $\sigma_1(n,\phi)\sigma_2(n,\phi)$, etc. This construction holds for functions that are not necessarily translation invariant.

7. Hilbert-Schmidt operators

Definition 7.1 (Hilbert-Schmidt operator). Let X be a complex and separable Hilbert space in which the norm is denoted by $\|\cdot\|$. A bounded linear operator on X is said to be a Hilbert-Schmidt operator if and only if there exists an orthonormal basis $\{\tau_m\}_{m=1}^{\infty}$ for X such that $\sum\limits_{m=1}^{\infty}\|A\tau_m\|_X^2$ is finite. If $A : X \to X$ is a Hilbert-Schmidt operator, then its norm is given by

$$\|A\|_{HS}^2 = \sum_{m=1}^{\infty}\|A\tau_m\|_X^2, \tag{96}$$

where $\{\tau_m\}_{m=1}^{\infty}$ is any orthonormal basis for X.

Theorem 7.2. *The pseudo-differential operator $T_\sigma : L^2(\mathbb{Z}^N) \to L^2(\mathbb{Z}^N)$ is a Hilbert-Schmidt operator if and only if $\sigma\in L^2(\mathbb{Z}^N \times \mathbb{T}^N)$. Moreover, if $T_\sigma : L^2(\mathbb{Z}^N) \to L^2(\mathbb{Z}^N)$ is a Hilbert-Schmidt operator, then*

$$\|T_\sigma\|_{HS} = (2\pi)^{-1/2}\|\sigma\|_{L^2(\mathbb{Z}^N\times\mathbb{T}^N)}. \tag{97}$$

Proof: Let $\{\tau_m\}_{m\in\mathbb{Z}^N}$ be the standard orthonormal basis for $L^2(\mathbb{Z}^N)$ which is defined by

$$\tau_m(n) = \begin{cases} 1 \text{ if } & n = m, \\ 0 & \text{otherwise.} \end{cases} \tag{98}$$

Using (91), the Fourier transform of τ_m for all $m \in \mathbb{Z}^N$ gives;

$$(\mathcal{F}_{\mathbb{Z}^N}\tau_m)(\phi) = e^{-im\cdot\phi}. \tag{99}$$

The pseudo-differential operator T_σ is

$$(T_\sigma\tau_m)(n) = \frac{1}{2\pi}\int_{\mathbb{T}^N} e^{in\cdot\phi}\sigma(n,\phi)(\mathcal{F}_{\mathbb{Z}^N}\tau_m)(\phi)d\phi \tag{100}$$

Plugging 99 into the previous gives

$$(T_\sigma\tau_m)(n) = \frac{1}{2\pi}\int_{\mathbb{T}^N} e^{in\cdot\phi}\sigma(n,\phi)e^{-im\cdot\phi}d\phi \tag{101}$$

$$= \frac{1}{2\pi}\int_{\mathbb{T}^N} e^{i(n-m)\cdot\phi}\sigma(n,\phi)d\phi \tag{102}$$

$$= \frac{1}{2\pi}\int_{\mathbb{T}^N} e^{-i(m-n)\cdot\phi}\sigma(n,\phi)d\phi, \tag{103}$$

Since $\frac{1}{2\pi}\int_{\mathbb{T}^N} e^{-i(m-n)\phi}\sigma(n,\phi)d\phi$ is the Fourier transform of the symbol on the Torus, thus

$$(T_\sigma\tau_m)(n) = (\mathcal{F}_{\mathbb{T}^N}\sigma)(n, m-n). \tag{104}$$

The fact that $T_\sigma\tau_m$ belongs to $L^2(\mathbb{Z}^N)$ implies its norm gives

$$\|T_\sigma\tau_m\|^2_{L^2(\mathbb{Z}^N)} = \sum_{n\in\mathbb{Z}^N} |(\mathcal{F}_{\mathbb{T}^N}\sigma)(n, m-n)|^2. \tag{105}$$

By definition, if T_σ is a Hilbert-Schmidt operator then its norm is given by

$$\|T_\sigma\|^2_{HS} = \sum_{m\in\mathbb{Z}^N} \|T_\sigma\tau_m\|^2_{L^2(\mathbb{Z}^N)}, \tag{106}$$

Substituting Eq. (105) into (106) we have

$$\|T_\sigma\|^2_{HS} = \sum_{m\in\mathbb{Z}^N}\sum_{n\in\mathbb{Z}^N} |(\mathcal{F}_{\mathbb{T}^N}\sigma)(n, m-n)|^2 \tag{107}$$

$$= \sum_{n\in\mathbb{Z}^N}\sum_{m\in\mathbb{Z}^N} |(\mathcal{F}_{\mathbb{T}^N}\sigma)(n, m-n)|^2 \qquad \left(\text{by Fubinis's Theorem}\right) \tag{108}$$

$$= \sum_{n\in\mathbb{Z}^N}\sum_{m\in\mathbb{Z}^N} |(\mathcal{F}_{\mathbb{T}^N}\sigma)(n, m)|^2 \qquad (\text{by translation invariant}) \tag{109}$$

$$= \frac{1}{2\pi}\sum_{n\in\mathbb{Z}^N}\int_{\mathbb{T}^N} |\sigma(n,\phi)|^2 d\phi \qquad (\text{by Plancherel formula}) \tag{110}$$

$$= \frac{1}{2\pi}\|\sigma\|^2_{L^2(\mathbb{Z}^N\times\mathbb{T}^N)}. \tag{111}$$

Hence, this finishes the proof.

8. L^p-Boundedness of ΨDO

Theorem 8.1. *Let σ be a measurable function on $\mathbb{Z}^N \times \mathbb{T}^N$ such that there exist a function $\omega \in L^2(\mathbb{Z}^N)$ for which*

$$|\sigma(n,\phi)| \leq |\omega(n)| \tag{112}$$

for all $n \in \mathbb{Z}^N$ and almost all $\phi \in [-\pi, \pi]$. Then $T_\sigma : L^2(\mathbb{Z}^N) \to L^2(\mathbb{Z}^N)$ is a bounded linear operator. Furthermore,

$$\|T_\sigma\|_{B(L^2(\mathbb{Z}^N))} \leq \frac{1}{\sqrt{2\pi}} \|\omega\|_{L^2(\mathbb{Z}^N)}, \tag{113}$$

where, $\|T_\sigma\|_{B(L^2(\mathbb{Z}^N))}$ is the norm of the bounded linear operator $T_\sigma : L^2(\mathbb{Z}^N) \to L^2(\mathbb{Z}^N)$.

Proof: By definition the Pseudo-differential operator is given by

$$(T_\sigma b)(n) = \frac{1}{2\pi} \int_{\mathbb{T}^N} e^{in\phi} \sigma(n,\phi) (\mathcal{F}_{\mathbb{Z}^N} b)(\phi) d\phi \tag{114}$$

with $f \in \mathcal{S}(\mathbb{Z}^N)$, taking the absolute square of both sides

$$|T_\sigma b|^2 = \left| \frac{1}{2\pi} \int_{\mathbb{T}^N} e^{in\phi} \sigma(n,\phi) (\mathcal{F}_{\mathbb{Z}^N} b)(\phi) d\phi \right|^2 \tag{115}$$

we know that

$$\sum_{n \in L^2(\mathbb{Z}^N)} |x|^2 = \|x\|^2 \tag{116}$$

This implies that

$$\|T_\sigma b\|^2 = \frac{1}{4\pi^2} \sum_{n \in L^2(\mathbb{Z}^N)} \left| \int_{\mathbb{T}^N} e^{in\phi} \sigma(n,\phi) (\mathcal{F}_{\mathbb{Z}^N} b)(\phi) d\phi \right|^2 \tag{117}$$

$$\leq \frac{1}{4\pi^2} \sum_{n \in L^2(\mathbb{Z}^N)} \int_{\mathbb{T}^N} |\sigma(n,\phi)|^2 |(\mathcal{F}_{\mathbb{Z}^N} b)(\phi)|^2 d\phi \qquad \text{(by Cauchy-Schwartz inequality)}. \tag{118}$$

Given that $|\sigma(n,\phi)| \leq |\omega(n)|$ thus we have

$$\|T_\sigma b\|^2_{L^2(\mathbb{Z}^N)} \leq \frac{1}{4\pi^2} \sum_{n \in L^2(\mathbb{Z}^N)} |\omega(n)|^2 \int_{\mathbb{T}^N} |(\mathcal{F}_{\mathbb{Z}^N} b)(\phi)|^2 d\phi \tag{119}$$

$$= \frac{1}{4\pi^2} \|\omega\|^2 \int_{\mathbb{T}^N} |(\mathcal{F}_{\mathbb{Z}^N} b)(\phi)|^2 d\phi. \tag{120}$$

By Plancheral's formula

$$\sum_{n\in\mathbb{Z}^N}|b(n)|^2=\frac{1}{2\pi}\int_{\mathbb{T}^N}\left|(\mathcal{F}_{\mathbb{Z}^N}b)(\phi)\right|^2d\phi. \tag{121}$$

This implies that

$$\|T_\sigma b\|^2=\frac{1}{4\pi^2}(2\pi)\|\omega\|^2\sum_{n\in\mathbb{Z}^N}|b(n)|^2 \tag{122}$$

$$=\frac{1}{2\pi}\|\omega\|^2_{L^2(\mathbb{Z}^N)}\|b\|^2_{L^2(\mathbb{Z}^N)}. \tag{123}$$

Therefore we have,

$$\|T_\sigma\|_{B(L^2(\mathbb{Z}^N))}\leq\frac{1}{2\pi}\|\omega\|_{L^2(\mathbb{Z}^N)} \tag{124}$$

Definition 8.2 (Fourier Approximation). Referring to the regular Fourier transform on the torus, for a given function σ,

$$(\mathcal{F}_{\mathbb{T}^n})\sigma(n,k)=\frac{1}{2\pi}\int_{\mathbb{T}^n}\sigma(n,\phi)e^{-ik\cdot\phi}d\phi \tag{125}$$

where: $k\in\mathbb{Z}^n$ is the frequency index, $k\cdot\phi=\sum_{j=1}^n k_j\phi_j$ is the dot product of k and ϕ, which represents the frequency of the exponential basis function, we can compute $(\mathcal{F}_{\mathbb{T}^n})\sim\sigma(n,k)$ using the expression

$$(\mathcal{F}_{\mathbb{T}^n})\sim\sigma(n,-k)=\frac{1}{2\pi}\int_{\mathbb{T}^n}\sigma(n,\phi)e^{-i(-k\cdot\phi)}d\phi \tag{126}$$

$$=\frac{1}{2\pi}\int_{\mathbb{T}^n}\sigma(n,\phi)\overline{e^{-i(k\cdot\phi)}}d\phi \tag{127}$$

$$=\overline{(\mathcal{F}_{\mathbb{T}^n})\sigma(n,k)}. \tag{128}$$

Then at a particular frequency $-k$, we obtain

$$(\mathcal{F}_{\mathbb{T}^n})\sim\sigma(n,n-k)=\overline{(\mathcal{F}_{\mathbb{T}^n})\sigma(n,k-n)}. \tag{129}$$

Remark 8.3. • The approximate Fourier transform at frequency $-k$ is simply the complex conjugate of the regular Fourier transform at frequency k.

- The approximate Fourier transform $(\mathcal{F}_{\mathbb{T}^n})\sim\sigma(n,-k)$ gives the complex conjugate of the Fourier coefficient at k, effectively transforming the function at a "negative" frequency $-k$.

- The $(\mathcal{F}_{\mathbb{T}^n})\sim\sigma(n,n-k)$ is a form of Fourier approximation or Fourier transform on the torus, where the function $\sigma(n,k)$ is transformed with respect to a shifted frequency $n-k$.

Theorem 8.4. *Consider σ a measurable function on $\mathbb{Z}^N \times \mathbb{T}^N$ such that we can find a positive constant C and a function $\omega \in L^1(\mathbb{Z}^N)$ for which*

$$|(\mathcal{F}_{\mathbb{T}^N}\sigma)(n,k)| \leq C|\omega(k)|, \qquad k,n \in \mathbb{Z}^N \tag{130}$$

Then $T_\sigma : L^p(\mathbb{Z}^N) \to L^p(\mathbb{Z}^N)$ is a bounded linear operator. Furthermore,

$$\|T_\sigma\|_{B(L^p(\mathbb{Z}^N))} \leq C\|\omega\|_{L^1(\mathbb{Z}^N)}, \tag{131}$$

where $\|T_\sigma\|_{B(L^p(\mathbb{Z}^N))}$ is the norm of the bounded linear operator T_σ on $L^p(\mathbb{Z}^N)$.

Proof: By definition the Fourier transform $\mathcal{F}_{\mathbb{Z}^N}b$ of $b \in L^1(\mathbb{Z}^N)$ is a function on the torus $\mathbb{T}^N$ which is expressed as

$$(\mathcal{F}_{\mathbb{Z}^N}b)(\phi) = \sum_{k \in \mathbb{Z}^N} b(n)e^{-ik\phi}, \qquad \phi \in \mathbb{T}^N \tag{132}$$

Let $b \in L^1(\mathbb{Z}^N)$. Then, T_σ (pseudo-differential operator) is defined as

$$(T_\sigma b)(n) = \frac{1}{2\pi}\int_{\mathbb{T}^N} e^{in\phi}\sigma(n,\phi)(\mathcal{F}_{\mathbb{Z}^N}b)(\phi)d\phi, \qquad n \in \mathbb{Z}^N. \tag{133}$$

Plugging Eq. (132) into (133) results

$$(T_\sigma b)(n) = \frac{1}{2\pi}\int_{\mathbb{T}^N} e^{in\phi}\sigma(n,\phi)\left(\sum_{k \in \mathbb{Z}^N} b(k)e^{-ik\phi}\right)d\phi \tag{134}$$

$$= \frac{1}{2\pi}\sum_{k \in \mathbb{Z}^N} b(k)\int_{\mathbb{T}^N} e^{-i(k-n)\phi}\sigma(n,\phi)d\phi \tag{135}$$

$$= \sum_{k \in \mathbb{Z}^N} b(k)(\mathcal{F}_{\mathbb{T}^N}\sigma)(n,k-n) \tag{136}$$

By definition (8.2), the Fourier approximation $(\mathcal{F}_{\mathbb{T}^N}\sigma) \sim$ of a function is given by

$$(\mathcal{F}_{\mathbb{T}^N}\sigma) \sim (n,-k) = (\mathcal{F}_{\mathbb{T}^N}\sigma)(n,k) \tag{137}$$

Thus

$$(T_\sigma b)(n) = \sum_{k \in \mathbb{Z}^N} b(k)(\mathcal{F}_{\mathbb{T}^N}\sigma)^{\sim}(n,n-k) \tag{138}$$

$$= ((\mathcal{F}_{\mathbb{T}^N}\sigma) \sim (n,\cdot) * b)(n) \qquad \text{(by convolution formula on } \mathbb{Z}^N\text{)}. \tag{139}$$

Taking absolute to the power p on both sides;

$$|(T_\sigma b)(n)|^p = |((\mathcal{F}_{\mathbb{T}^N}\sigma) \sim (n,\cdot) * b)(n)|^p \tag{140}$$

$$\sum_{n \in \mathbb{Z}^N} |(T_\sigma b)(n)|^p = \sum_{n \in \mathbb{Z}^N} |((\mathcal{F}_{\mathbb{T}^N}\sigma) \sim (n,\cdot) * b)(n)|^p \tag{141}$$

$$\leq \sum_{n\in\mathbb{Z}^N} ((|(\mathcal{F}_{\mathbb{T}^N}\sigma) \sim (n,\cdot)| * |b|)(n))^p \qquad \text{(by Cauchy-Schwartz inequality)}. \tag{142}$$

We know that $|(\mathcal{F}_{\mathbb{T}^N}\sigma) \sim (n,\cdot)| \ \leq C|\omega|$ (thus)

$$\sum_{n\in\mathbb{Z}^N} |(T_\sigma b)(n)|^p \leq \sum_{n\in\mathbb{Z}^N} ((C|\omega| * |b|)(n))^p \tag{143}$$

$$= C^p \sum_{n\in\mathbb{Z}^N} ((|\omega| * |b|)(n))^p. \tag{144}$$

By Young's Inequality for convolution we have

$$\|T_\sigma b\|^p_{L^p(\mathbb{Z}^N)} \leq C^p \|\omega\|^p_{L^1(\mathbb{Z}^N)} \|b\|^p_{L^p(\mathbb{Z}^N)}. \tag{145}$$

The fact that $L^1(\mathbb{Z}^N)$ in $L^p(\mathbb{Z}^N)$ completes the boundedness of T_σ in $L^p(\mathbb{Z}^N)$ for $1<p<\infty$.

9. L^p – compactness of ΨDO

Here, we give another condition on the symbol function σ which guarantees compactness of pseudo-differential operator.

Theorem 9.1. *Consider σ a measurable function on $\mathbb{Z}^N \times \mathbb{T}^N$ such that we can find a positive function C on $\mathbb{Z}^N$ and a function $\omega \in L^1(\mathbb{Z}^N)$ for which,*

$$|(\mathcal{F}_{\mathbb{T}^N}\sigma)(n,m)| \ \leq C(n)|\omega(m)|, \quad m,n\in\mathbb{Z}^N, \tag{146}$$

and

$$\lim_{|n|\to\infty} C(n) = 0. \tag{147}$$

Then the pseudo-differential operator $T_\sigma : L^p(\mathbb{Z}^N) \to L^p(\mathbb{Z}^N)$ is a compact operator for $1\leq p<\infty$.

Proof: We consider the sequence $(T_{\sigma_N})_{N>0}$ (where)

$$\sigma_N(n,\phi) = \begin{cases} \sigma(n,\phi), |n| \ \leq N, \\ 0, |n| \ > N. \end{cases} \tag{148}$$

Referring to the definition (94), and for any $f \in L^1(\mathbb{Z}^N)$ we have

$$(T_{\sigma_N})(b)(n) = \begin{cases} \dfrac{1}{2\pi}\displaystyle\int_{\mathbb{T}^N} e^{in\cdot\phi}\sigma(n,\phi)(\mathcal{F}_{\mathbb{Z}^N}b)(\theta)d\theta, |n| \ \leq N, \\ 0, |n| \ > N. \end{cases} \tag{149}$$

Substructing T_{σ_N} form T_σ (see (94)) gives

$$(T_\sigma - T_{\sigma_N})(b)(n) = \frac{1}{2\pi}\int_{\mathbb{T}^N} e^{in\cdot\phi}(\sigma - \sigma_N)(n,\phi)(\mathcal{F}_{\mathbb{Z}^N} b)(\theta)d\theta. \tag{150}$$

Substituting (91) into (150) gives

$$(T_\sigma - T_{\sigma_N})(b)(n) = \frac{1}{2\pi}\sum_{n\in\mathbb{Z}^N} b(n)\int_{\mathbb{T}^N} e^{-i(m-n)\cdot\phi}(\sigma - \sigma_N)(n,\phi)d\phi. \tag{151}$$

Then

$$(T_\sigma - T_{\sigma_N})(f)(n) = \sum_{m\in\mathbb{Z}^N} f(m)(\mathcal{F}_{\mathbb{T}^N}(\sigma - \sigma_N))(n, m-n). \tag{152}$$

At this stage, we use Fourier transformation equivalent, that is,

$$(\mathcal{F}_{\mathbb{T}^N}(\sigma - \sigma_N))(n, m-n) = (\mathcal{F}_{\mathbb{T}^N}(\sigma - \sigma_N))^{\sim}(n, n-m), \tag{153}$$

hence we have

$$(T_\sigma - T_{\sigma_N})(b)(n) = \sum_{m\in\mathbb{Z}^N} b(m)(\mathcal{F}_{\mathbb{T}^N}(\sigma - \sigma_N))^{\sim}(n, n-m). \tag{154}$$

Referring to the convolution formula on $\mathbb{Z}^N$, we write

$$(T_\sigma - T_{\sigma_N})(b)(n) = ((\mathcal{F}_{\mathbb{T}^N}(\sigma - \sigma_N))^{\sim}(n,\cdot) * b)(n). \tag{155}$$

Taking the absolute value of both sides to the power p (gives)

$$|(T_\sigma - T_{\sigma_N})(b)(n)|^p = \left|((\mathcal{F}_{\mathbb{T}^N}(\sigma - \sigma_N))^{\sim}(n,\cdot) * b)(n)\right|^p \tag{156}$$

$$\leq ((|(\mathcal{F}_{\mathbb{T}^N}(\sigma - \sigma_N))^{\sim}(n,\cdot)| * |b|)(n))^p \tag{157}$$

Taking sum on both sides for $n\in\mathbb{Z}^N$ (gives)

$$\|(T_\sigma - T_{\sigma_N})(b)\|^p_{L^p(\mathbb{Z}^N)} \leq \sum_{n\in\mathbb{Z}^N} ((|(\mathcal{F}_{\mathbb{T}^N}(\sigma - \sigma_N))^{\sim}(n,\cdot)| * |b|)(n))^p \tag{158}$$

$$= \sum_{|n|\ >N} ((|(\mathcal{F}_{\mathbb{T}^N}\sigma)^{\sim}(n,\cdot)| * |b|)(n))^p \quad (\sigma_N = 0 \text{ for } |n|\ >N), \tag{159}$$

By hypothesis, we have

$$|(\mathcal{F}_{\mathbb{T}^N\sigma})(n,m)|\ \leq C(n)|\omega(m)|. \tag{160}$$

and $\lim_{|n|\to\infty} C(n) = 0$ means that for all $\varepsilon > 0$ there exists $N_0 > 0$ such that $|C(n)|\ < \varepsilon$, for all $n > N_0$. This implies that

$$|C(n)| \leq \varepsilon, \tag{161}$$

then

$$\left|(\mathcal{F}_{\mathbb{T}^N\sigma})^{\sim}(n,\cdot)\right|^p \leq \varepsilon^p|\omega|^p. \tag{162}$$

Hence,

$$\|(T_\sigma - T_{\sigma_N})(b)\|^p_{L^p(\mathbb{Z}^N)} \leq \sum_{|n|>N} ((\varepsilon|\omega| * |b|)(n))^p \tag{163}$$

$$\leq \varepsilon^p \sum_{|n|>N} ((|\omega| * |b|)(n))^p \tag{164}$$

$$= \varepsilon^p \|\omega * b\|^p_{L^p(\mathbb{Z}^N)} \tag{165}$$

$$\leq \varepsilon^p \|\omega\|^p_{L^1(\mathbb{Z}^N)} \|b\|^p_{L^p(\mathbb{Z}^N)} \tag{166}$$

this is equivalent to

$$\|T_\sigma - T_{\sigma N}\|_{B(L^p(\mathbb{Z}^N))} \leq \varepsilon \|\omega\|_{L^1(\mathbb{Z}^N)}. \tag{167}$$

From our results, it implies that T_σ is the limit in norm of a sequence of compact operator on $L^p(\mathbb{Z}^N)$ hence T_σ is L^p-compact.

10. Conclusion

The extension of the result investigated in Ref. [1] have been achieved. By imposing necessary and sufficient conditions on the symbol $\sigma \in S^m(\mathbb{T}^N \times \mathbb{Z}^N)$ we proved the corresponding pseudo-differential operators are Hilbert-Schmidt operators, boundedness and compactness on $L^p(\mathbb{Z}^N)$ for $1 \leq p < \infty$. By changing the space and studying pseudo-differential operators conserve their properties.

Regarding the behavior of this kind of operators, one can envisage to study them on other topological spaces and investigate the L^p-nuclearity, the composition formula of pseudo-differential operators, the amplitude and Kernel representation, the transpose and the adjoint formula of pseudo-differential operators.

Acknowledgements

I am grateful to IntechOpen for the opportunity to contribute to this current book project, Operator Theory; Recent Developments and Applications. This is a golden opportunity for my professional career, especially in terms of research endeavors.

Thanks

Many thanks to Dr. Linda Naa Adjeley Botchway, my research partner, for the support and advice throughout this work.

A. Appendix

In this Appendix, an important tool in the theory of pseudo-differential operators is presented.

The theory of pseudo-differential operators is not only discussed and investigated to a particular class of operators but to general linear continuous operators on the space. In fact, given an operator A such as $A : C^{\infty}(\mathbb{T}^N) \to C^{\infty}(\mathbb{T}^N)$ is a continuous linear operator, then it can be shown that A can be written in the form $A = \mathrm{Op}(a)$ with the symbol $\sigma = a(x,k)$ defined by

$$a(x,k) := e_{-k}(x) A e_k(x) = e^{-i2\pi x\cdot k} A\left(e^{i2\pi x\cdot k}\right), \tag{168}$$

where $e_k(x) = e^{i2\pi x\cdot k}$ for all $k \in \mathbb{Z}^N$ and $x \in \mathbb{T}^n$.

By acting the operator A on f using the Fourier inversion formula [12], we obtain

$$Af(x) = A\left(\sum_{k\in\mathbb{Z}^N} e^{i2\pi x\cdot k}\hat{f}(k)\right) \tag{169}$$

$$= \sum_{k\in\mathbb{Z}^N} A\left(e^{i2\pi x\cdot k}\right)\hat{f}(k) \tag{170}$$

$$= \sum_{k\in\mathbb{Z}^N} e^{2\pi i x\cdot k}\sigma(k,x)\hat{f}(k) = \mathrm{Op}(a)f(x). \tag{171}$$

This implies that any linear operator is a pseudo-differential operators. This statement motivates the study of pseudo-differential operators. The proposition below allows us to checking the above statement we just claimed.

Proposition 10.1. Let A be a pseudo-difference operator. Then its symbol is given by

$$\sigma(x,k) = e^{-i2\pi x\cdot k} A e_k(x), \tag{172}$$

where $e_k(x) = e^{i2\pi x\cdot k}$, for all $k \in \mathbb{Z}^N$ and $x \in \mathbb{T}^N$.

Proof: For the function $e_t(y) = e^{i2\pi y\cdot t}$, its Fourier transform is given formally by

$$\hat{e}_t(x) = \int_{\mathbb{T}^N} e^{-i2\pi x\cdot t} e^{i2\pi y\cdot t}\,\mathrm{d}x. \tag{173}$$

Plugging this into the formula

$$\mathrm{Op}(\sigma)f(x) = \sum_{t\in\mathbb{Z}^N} e^{i2\pi x\cdot t}\sigma(x,t)\hat{f}(x), \tag{174}$$

it follows that

$$\mathrm{Op}(\sigma)e_k(y) = \sum_{t\in\mathbb{Z}^N}\int_{\mathbb{T}^N} e^{i2\pi x\cdot k}\sigma(x,k)e^{-i2\pi x\cdot t}e^{i2\pi y\cdot t}\,\mathrm{d}x \tag{175}$$

$$= \sum_{t\in\mathbb{Z}^N}\int_{\mathbb{T}^N} e^{-i2\pi x\cdot(t-k)}\sigma(x,k)e^{i2\pi y\cdot t}\,\mathrm{d}x \tag{176}$$

$$= \sum_{t\in\mathbb{Z}^n} \hat{\sigma}(t-k,k)e^{i2\pi y\cdot t} \tag{177}$$

$$= \sum_{m\in\mathbb{Z}^N} \hat{\sigma}(m,k)e^{i2\pi y\cdot m}e^{i2\pi y\cdot k} \qquad (\text{where } t-k=m) \tag{178}$$

$$= \sigma(y,k)e^{i2\pi y\cdot k}, \tag{179}$$

where $\hat{\sigma}$ is for the Fourier transform on $\mathbb{T}^N$ in the second variable.

Example 10.2 (Illustrated example). Here, we show that the Laplacian $\mathcal{L}_x$ which is an elliptic differential operator is a pseudo-differential operator on any space but we still consider the lattice case. One can use the similar approach to prove this on $\mathbb{R}^N$.

Consider the operator given by

$$f\mapsto\mathcal{L}_x f = \sum_{i=1}^{N}\frac{\partial^2}{\partial x_i^2}f. \tag{180}$$

Let f be a plane wave such as, $f(x) = e^{i2\pi x\cdot k}$, then there exists a simple relationship

$$\mathcal{L}_x\left(e^{i2\pi x\cdot k}\right) = -4\pi^2|k|^2e^{i2\pi x\cdot k}, \tag{181}$$

where $|k|^2 = k_1 + \ldots + k_n$ and $x = (x_1, \ldots, x_N)$.

Now, we arrive at the equation

$$\mathcal{L}_x f(x) = \mathcal{L}_x\left(\sum_{k\in\mathbb{Z}^N} e^{-i2\pi x\cdot k}f(k)\right) \tag{182}$$

$$= \sum_{k\in\mathbb{Z}^N}\mathcal{L}_x\left(e^{-i2\pi x\cdot k}\right)f(k) \tag{183}$$

$$= \sum_{k\in\mathbb{Z}^N} e^{-i2\pi x\cdot k}\left(-4\pi^2|k|^2\right)f(k). \tag{184}$$

By this, it is clear to see that the Laplacian is also a pseudo-differential operator of symbol $-4\pi^2|k|^2$. Then, we claim that every linear differential operator with smooth and bounded coefficients is a pseudo-differential operator.

Then for any $\mu\in\mathbb{R}$, we can define the operators $(1-\mathcal{L}_x)^\mu$ as pseudo-differential operators with symbol $\left(1+4\pi^2|k|^2\right)^{\mu/2}$.

Abbreviations

$a\in S^m$	symbol
P	differential operator
Ψ DO	pseudo-differential operator
$\tau_m(n)$	orthonormal basis
$\mathbb{Z}$	set of integers
$\mathbb{Z}^N$	lattice
$\mathcal{S}$	Schwartz space
L^p	Lebesgue space
α, ψ	multi-indices
$\mathcal{F}_\mathbb{Z}$	Fourier transform on $\mathbb{Z}$

$D^\alpha = \partial_x^\alpha$	derivatives
C_α	constant
$\mathbb{S}^1$	unit circle
$\mathbb{T}^N$	N-dimensional torus

Author details

Perrin G. Kibiti Pembe[1*] and Linda N.A. Botchway[2]

1 African Institute for Mathematical Sciences (AIMS), Ghana

2 University of Ghana (UG), Legon, Ghana

*Address all correspondence to: gael@aims.edu.gh

References

[1] Šubin MA. Pseudodifferential Operators and Spectral Theory. Soviet Mathematics Series. Springer-Verlag; 1987. Available from: https://books.google.fr/books?id=bzrvAAAAMAAJ. ISBN: 9783540136217, ICCN: lc86010213

[2] Seeley R. Pseudo-differential operators. In: Topics in Pseudo-Differential Operators. Springer; 2011. pp. 167-305

[3] Adler M. On a trace functional for formal pseudo-differential operators and the symplectic structure of the Korteweg-devries type equations. Inventiones Mathematicae. 1978/79;**50**: 219-248. Available from: http://eudml.org/doc/142614

[4] Ruzhansky M, Turunen V. On pseudo-differential operators on group SU(2). 2008. Available from: https://arxiv.org/abs/0802.2780. eprint 0802.2780

[5] Ruzhansky M, Turunen V. Pseudo-Differential Operators and Symmetries: Background Analysis and Advanced Topics. Vol. 2. Basel, Switzerland: Birkhäuser; 2010. DOI: 10.1007/978-3-7643-8514-9

[6] Taylor ME. Pseudo-Differential Operators and Nonlinear PDE, Vol. 100 of Progress in Mathematics. Vol. 81. Boston, MA: Birkhäuser Boston Inc.; 1991

[7] Cardona D. Nuclear pseudo-differential operators in Besov spaces on compact lie groups. 2016. Available from: https://arxiv.org/abs/1610.09042. eprint 1610.09042

[8] Cardona D. On the boundedness of periodic pseudo-differential operators. arXiv e-prints. Jan 2017: arXiv:1701.08184. DOI: 10.48550/arXiv.1701.08184. Available from: https://ui.adsabs.harvard.edu/abs/2017arXiv170108184C

[9] Molahajloo S, Wong WM. Pseudo-Differential Operators on S^1. Springer; 2008. pp. 297-306

[10] Molahajloo S. Pseudo-differential operators on Z. In: Pseudo-Differential Operators: Complex Analysis and Partial Differential Equations. Vol. 18. Nombre: 2. 2009. Springer; 2008. pp. 313-221

[11] Wong MW. An Introduction to Pseudo-Differential Operators. Vol. 6. World Scientific Publishing Co Inc; 2008

[12] Brezis H. Functional Analysis, Sobolev Spaces and Partial Differential Equations. Springer Science and Business Media; 2010

Chapter 4

On the Generalized Quantum Linear Momentum Operator

Jesús García-Ravelo, Jesús García-Martínez, Jesús Morales and José Juan Peña

Abstract

From the canonical form of the position-dependent mass Hamiltonian, the generalized quantum linear momentum operator (GLMO) is obtained. Such operator is straightforwardly generated by rewritten the Schrödinger Hamiltonian in terms of von Ross kinetic energy operator, expressed as the square of the Hermitian GLMO plus a potential energy function. In this scheme, the hermiticity property and the arbitrary ambiguity parameters remain in force. In addition, by means of different methods, it is shown that the GLMO has a unique structure regardless of the form of the mass distribution. Hence, the proposed approach is valid for any physically plausible position-dependent mass distribution $m(x)$. As an example of useful applications of our proposal, two specific $m(x)$ with arbitrary ambiguity parameters are considered. The application is given for some potential models, such as the null potential, the harmonic oscillator, the double well potential, and the Coulomb-like potential. For some particular values of the ambiguity parameters, our results agree with some already published in the literature, which means that our method can be considered as an improvement vis-a-vis those proposals on the same subject given until now.

Keywords: effective mass Schrödinger equation, position-dependent mass, linear momentum operator, von-Roos position-dependent mass Hamiltonian, Hermitian operator

1. Introduction

The quantum mechanics operators are crucial elements that, acting on the Hamiltonian and the wave functions, define some characteristics of the system under study. Regarding the position and linear momentum operators, linearity, non-additivity, Hermiticity, commutation relationships, or algebraic groups are only some of their properties through which one can know about the nature of the system. In the case of the position-dependent mass Schrödinger equation, certain properties like boundary conditions [1], commutation relations [2], and generalized kinetic energy operators [3], have been proposed. In this scenario, motivated by the non-extensive statistical mechanics [4, 5], the concept of deformed quantum mechanics [6] has been related with a change in the linear momentum operator [7, 8]. Similar results on a position-dependent infinitesimal translation operator are given by Costa Filho et al. [9] and by Mazharimousavi [10].

Consequently, a new deformed Hermitian operator leads, among other results, to a deformation of the canonical commutation relations [11] as for example, the proposal of a minimal length uncertainty for quantum systems in the momentum space gives rise to a modification of the commutation relation for the position and linear momentum operators [12]. This new representation for the linear momentum operator demands to reformulate the structure of the Hamiltonian, as well as all quantities related to it. In this work, with the purpose to be in agreement with similar studies, the establishment of the Generalized Quantum Linear Momentum Operator (GLMO) is related to the quantum linear momentum operator for the position-dependent mass Schrödinger equation with this aim, we are proposing to obtain this GLMO straightforwardly by using simple commutation relations after transforming the Hamiltonian with position-dependent mass into its canonical form, that is, with the kinetic energy operator given in terms of a square Hermitian operator plus a potential energy function. As will be seen next, our proposition has two advantages: one, it is valid to any position-dependent mass distribution, and two, it can be applied to whichever ambiguity parameter. Furthermore, by using different methods, in the following we have shown that this GLMO is unique. Also, as a useful application of our proposal, we consider as examples, two particular position-dependent mass distribution in order to solve some quantum potentials in the frame of the von-Ross Hamiltonian for arbitrary ambiguity parameters [13] as well as to show how our approach gives rise to specific results already appeared in the literature.

2. On the generalization of the quantum linear momentum operator

The position-dependent mass Schrödinger equation, written in terms of the von Roos kinetic energy operator $T_{vR}(x)$, is given by

$$[T_{vR}(x) + V(x)]\psi_n(u) = E_n\psi_n(x) \tag{1}$$

where

$$T_{vR}(x) = -\frac{\hbar^2}{4}\left(m^\alpha(x)\frac{d}{dx}m^\beta(x)\frac{d}{dx}m^\gamma(x) + m^\gamma(x)\frac{d}{dx}m^\beta(x)\frac{d}{dx}m^\alpha(x)\right) \tag{2}$$

that, according to the value of the ambiguity parameters $\alpha + \beta + \gamma = -1$, may adopt different hermitian forms [14–16]. The Eq. (1) can be rewritten as [17].

$$\left[-\frac{d}{dx}\left(\frac{\hbar^2}{2m(x)}\right)\frac{d}{dx} + U(x)\right]\psi_n(x) = E_n\psi_n(x) \tag{3}$$

with

$$U(x) = V(x) + \frac{\hbar^2}{4}(\beta+1)\frac{m''(x)}{m^2(x)} - \frac{\hbar^2}{2}[\alpha(\alpha+\beta+1) + \beta + 1]\frac{(m'(x))^2}{m^3(x)} \tag{4}$$

where the particular case $\beta = -1, \alpha = \gamma = 0$ leads to $U(x) = V(x)$ [18].

Then, with the aim of finding the GLMO the von Ross Hamiltonian should be expressed into a canonical form $H = \frac{\hat{p}^2}{2m_0} + u$. For that purpose, we start by rewriting the Hamiltonian in Eq. (3) as

DOI: http://dx.doi.org/10.5772/intechopen.1009028

$$\hat{H} = \frac{d}{dx}\left(\frac{i\hbar}{\sqrt{2m(x)}}\right)\left(\frac{i\hbar}{\sqrt{2m(x)}}\right)\frac{d}{dx} + U(x) \tag{5}$$

that, with the aid of the commutator

$$\left[\frac{d}{dx}, \frac{i\hbar}{\sqrt{2m(x)}}\right] = i\hbar\left(\frac{1}{\sqrt{2m(x)}}\right)' \tag{6}$$

leads to

$$\hat{H} = \left[\frac{i\hbar}{\sqrt{2m(x)}}\frac{d}{dx} + i\hbar\left(\frac{1}{\sqrt{2m(x)}}\right)'\right]\left[\frac{i\hbar}{\sqrt{2m(x)}}\frac{d}{dx}\right] + U(x). \tag{7}$$

Thus, after some operational calculations one gets

$$\hat{H} = \left(\frac{i\hbar}{\sqrt{2m(x)}}\frac{d}{dx}\right)^2 - 2\frac{i\hbar m'(x)}{4m(x)\sqrt{2m(x)}}\left(\frac{i\hbar}{\sqrt{2m(x)}}\frac{d}{dx}\right) + U(x), \tag{8}$$

$$\hat{H} = \left(\frac{i\hbar}{\sqrt{2m(x)}}\frac{d}{dx}\right)^2 - \frac{i\hbar m'(x)}{4m(x)\sqrt{2m(x)}}\left(\frac{i\hbar}{\sqrt{2m(x)}}\frac{d}{dx}\right) - \frac{i\hbar m'(x)}{4m(x)\sqrt{2m(x)}}\left(\frac{i\hbar}{\sqrt{2m(x)}}\frac{d}{dx}\right) + U(x). \tag{9}$$

Finally, by using the commutator

$$\left[\frac{i\hbar}{\sqrt{2m(x)}}\frac{d}{dx}, Q(x)\right] = i\hbar\frac{Q'(x)}{\sqrt{2m(x)}}, \tag{10}$$

with $Q(x) = \frac{i\hbar m'(x)}{4m(x)\sqrt{2m(x)}}$, Eq. (9) writes down as

$$\begin{aligned}\hat{H} = &\left(\frac{i\hbar}{\sqrt{2m(x)}}\frac{d}{dx}\right)^2 - \frac{i\hbar m'(x)}{4m(x)\sqrt{2m(x)}}\left(\frac{i\hbar}{\sqrt{2m(x)}}\frac{d}{dx}\right) \\ &-\left(\frac{i\hbar}{\sqrt{2m(x)}}\frac{d}{dx}\right)\left(\frac{i\hbar m'(x)}{4m(x)\sqrt{2m(x)}}\right) - \frac{\hbar^2\left(\frac{m'(x)}{4m(x)\sqrt{m(x)}}\right)'}{\sqrt{2m(x)}} + U(x),\end{aligned} \tag{11}$$

adopting the final form

$$\hat{H} = \left(-\frac{i\hbar}{\sqrt{2m(x)}}\frac{d}{dx} - \frac{i\hbar}{2}\left(\frac{1}{\sqrt{2m(x)}}\right)'\right)^2 + U(x) + \left[\frac{\hbar^2}{4}\left(\left(\frac{1}{\sqrt{2m(x)}}\right)'\right)^2 + \frac{\hbar^2}{2\sqrt{2m(x)}}\left(\frac{1}{\sqrt{2m(x)}}\right)''\right] \tag{12}$$

Consequently, from the above equation, it is possible to identify a canonical form of the von-Ross Hamiltonian with position-dependent mass as

$$\hat{H} = \frac{1}{2m_0}\hat{p}_{eff}^2 + u_{eff}(x), \tag{13}$$

where p_{eff} is the GLMO given by

$$\hat{p}_{eff} = -\frac{i\hbar}{\sqrt{M(x)}}\frac{d}{dx} - \frac{i\hbar}{2}\left(\frac{1}{\sqrt{M(x)}}\right)' \tag{14}$$

and $u_{eff}(x)$ is the *effective* potential

$$u_{eff}(x) = U(x) + \frac{\hbar^2}{2m_0}\left[\left(\left(\frac{1}{2\sqrt{M(x)}}\right)'\right)^2 + \frac{1}{\sqrt{M(x)}}\left(\frac{1}{2\sqrt{M(x)}}\right)''\right]. \tag{15}$$

that, after using Eq. (4), becomes

$$u_{eff}(x) = V(x) + \frac{\hbar^2}{4m_0}\left(\beta + \frac{1}{2}\right)\frac{M''(x)}{M^2(x)} - \frac{\hbar^2}{2m_0}\left[\alpha(\alpha+\beta+1) + \beta + \frac{9}{16}\right]\frac{(M'(x))^2}{M^3(x)}, \tag{16}$$

where $M(x) = m(x)/m_0$. At this point, it should be noted that the particular case of constant mass $m(x) = m_0$ ($M(x) = 1$) leads to

$$\hat{p}_{eff} = \hat{p} \text{ and } u_{eff}(x) = V(x) \tag{17}$$

where $\hat{p} = -i\hbar(d/(dx))$ is the standard linear momentum operator.

Let us see now the hermiticity of the GLMO. For that, let us consider a general operator $\hat{A}$ given by

$$\hat{A} = -ig(x)\frac{d}{dx} - if(x), \tag{18}$$

where $f(x)$ and $g(x)$ are two arbitrary real functions. In this case, the hermiticity condition for an arbitrary operator $\hat{\mathcal{O}}$

$$\int \left(\hat{\mathcal{O}}\psi\right)^* \psi\, dx) = \int \psi^* \hat{\mathcal{O}}\, \psi\, dx) \tag{19}$$

applied to $\hat{A}$, yields

$$\int \left(\hat{A}\,\psi\right)^* \psi\, dx) = \int \psi^* \hat{A}\,\psi\, dx) + i\int \psi^* (2f(x) - g'(x))\,\psi\, dx). \tag{20}$$

So, to have a Hermitian operator $\hat{A}$ in Eq. (18), both function $f(x)$ and $g(x)$ must meet the condition $2f(x) = g'(x)$. Hence, by comparing the operator $\hat{A}$ with $\hat{p}_{eff}$ of Eq. (14), the election of the function $g(x) = \hbar/\sqrt{M(x)}$ implies that $2f(x) = g'(x)$ leading to a Hermitian operator in Eq. (14). By the way, the operator $\hat{A}$ has a commutation relation

$$\left[x, \hat{A}\right] = ig(x) \tag{21}$$

whereas the generalized linear momentum operator satisfies

$$\left[x, \hat{p}_{eff}\right] = \frac{i\hbar}{\sqrt{M(x)}}, \tag{22}$$

as well as the generalized uncertainty relation is

$$\Delta x\, \Delta \hat{p}_{eff} \geq \frac{\hbar}{2\sqrt{M(x)}}, \tag{23}$$

The case of $M(x) = 1$ reproduces the already-known result $\Delta x \Delta \hat{p} \geq \frac{\hbar}{2}$.

3. Canonical commutation relations of the quantum dynamical variables

Once the von-Roos Hamiltonian has been written in a canonical form, without loss of generality we propose a deformation parameter λ such that the mass distribution is written down as $m(x, \lambda) = m_0 M(x, \lambda)$ where the particular case of constant mass is reached when

$$\lim_{\lambda \to 0} m(x, \lambda) = m_0 M(x, 0) = m_0, \tag{24}$$

Consequently, $M(x, 0) = 1$ and the λ-deformed canonical variables x_λ and p_λ are introduced as

$$x_\lambda = \int \sqrt{M(x, \lambda)}\, dx \tag{25}$$

and

$$p_\lambda = -\frac{i\hbar}{\sqrt{M(x, \lambda)}} \frac{d}{dx} - \frac{i\hbar}{2} \left(\frac{1}{\sqrt{M(x, \lambda)}} \right)' \tag{26}$$

where $p_\lambda = p_{eff}$ is the GLMO given in Eq. (14). Both x_λ and p_λ fulfill the standard form of the quantum commutator and Heisenberg's uncertainty principle

$$\left[x_\lambda, \hat{p}_\lambda\right] = i\hbar, \tag{27}$$

$$\Delta x_\lambda \Delta \hat{p}_\lambda \geq \frac{\hbar}{2}. \tag{28}$$

Hence, we have deduced a kind of generalized λ-deformed quantum mechanics within the frame of the position-dependent mass, in which the usual phase space (x, p) is mapped into new coordinates (x_λ, p_λ) preserving the same properties and quantum relations as the original variables. It is worth mentioning that these λ-deformed variables, x_λ and p_λ, are the generalized version of those proposed by da Costa et al. [6, 8], who have presented this type of λ-deformed quantum formalism for the case of the particular mass distributions $m(x,\lambda) = m_0(1+\lambda x)^{-2}$ and $m(x,\lambda) = m_0(1+\lambda^2 x^2)^{-1}$. Hence, in Section 4, we consider these position -dependent mass distributions in order to compare our proposal with da Costa results.

Next, we are going to show that the Hermitian GLMO has a unique representation. To that, by considering the general structure of the operator $\hat{A}$ given in Eq. (18) along with the hermiticity condition given in Eq. (20), one leads to

$$\hat{A} = -ig(x)\frac{d}{dx} - \frac{i}{2}g'(x), \tag{29}$$

where, regardless of the form of the function $g(x)$, the operator $\hat{A}$ will always be Hermitian, for example, to factorize into a canonical form the von Roos Hamiltonian operator is given in Eq. (13), it is necessary to use the specific function $g(x) = \frac{i\hbar}{\sqrt{M(x,\lambda)}}$ leading to

$$\hat{A} = -\frac{i\hbar}{\sqrt{M(x,\lambda)}}\frac{d}{dx} - \frac{i\hbar}{2}\left(\frac{1}{\sqrt{M(x,\lambda)}}\right)' = \hat{p}_{eff}, \tag{30}$$

which means the Hermitian GLMO has a unique representation of a given $M(x,\lambda)$. In this regard, it should be pointed out that there are different ways to get the GLMO although some of the already proposals have been derived, as mentioned, for particular cases of specific position-dependent mass distributions [6, 8]. So, in the most general case of arbitrary $M(x,\lambda)$ related to the λ-deformed differential operator, we consider the λ-deformed generalized exponential function $exp_{\lambda}(x)$ such that

$$\lim_{\lambda\to 0} exp_{\lambda}(x) = exp\,(x) \tag{31}$$

where $exp\,(x)$ stands for the standard exponential function. Thus, according with Eqs. (24) and (25)

$$\lim_{\lambda\to 0} x_{\lambda} = x, \tag{32}$$

then one can set

$$exp_{\lambda}(x) = exp\,(x_{\lambda}) \tag{33}$$

At this point, due to the fact that $\frac{d}{dx}exp\,(x) = exp\,(x)$, the λ-deformed differential operator that leaves invariant the generalized exponential function $exp_{\lambda}(x)$ will be $D_{\lambda} = \frac{1}{\sqrt{M(x,\lambda)}}\frac{d}{dx}$ such that

$$D_{\lambda}\, exp_{\lambda}(x) = \frac{1}{\sqrt{M(x,\lambda)}}\frac{d}{dx} exp\,(x_{\lambda}) = exp_{\lambda}(x) \tag{34}$$

where we have used Eqs. (25) and (33). Also, it is important to highlight that the above operator plays the role of the λ-deformed derivative operator in the new x_{λ} space. So, in this case, the corresponding non-Hermitian linear momentum operator will be $-i\hbar D_{\lambda}$ which can be set as a Hermitian operator by means of the procedure given above for the operator $\hat{A}$. Namely$-i\hbar D_\lambda$ transforms into a Hermitian operator by simply adding the factor $\frac{-i\hbar}{2}\left(\frac{1}{\sqrt{M(x,\lambda)}}\right)'$. That is

$$\hat{D}_\lambda = -\frac{i\hbar}{\sqrt{M(x,\lambda)}}\frac{d}{dx} - \frac{i\hbar}{2}\left(\frac{1}{\sqrt{M(x,\lambda)}}\right)' \tag{35}$$

corresponds to the p_{eff} GLMO obtained previously in Eq. (30).

Finally, let us see another method to find the GLMO. A straight forwardly way is using as an analogy the procedure used to obtain the Standard Linear Momentum Operator (SLMO) in quantum mechanics. In fact, according to the standard Schrödinger equation, the expectation value of the SLMO is written down as

$$\langle\hat{p}\rangle = m\frac{d}{dx}\langle x\rangle = \int \psi^*(x,t)\left(-i\hbar\frac{d}{dx}\right)\psi(x,t)\,dx \tag{36}$$

which, by analogy, the GLMO will be given by

$$\langle\hat{p}_\lambda\rangle = \int \varphi^*(x_\lambda,t)(-i\hbar D_\lambda)\varphi(x_\lambda,t)\,dx_\lambda \tag{37}$$

Thus, by considering the transformation $dx_\lambda = \sqrt{M(x,\lambda)}dx$ along with $\varphi(x_\lambda,t) = (M(x,\lambda))^{-\frac{1}{4}}\psi(x,t)$ one leads to

$$\begin{aligned}
\langle\hat{p}_\lambda\rangle &= \int (M(x,\lambda))^{-\frac{1}{4}}\psi^*(x,t)\left(-\frac{i\hbar}{\sqrt{M(x,\lambda)}}\frac{d}{dx}\right)(M(x,\lambda))^{-\frac{1}{4}}\psi(x,t)\sqrt{M(x,\lambda)}\,d \\
&= \int \psi^*(x,t)\left(-\frac{i\hbar}{\sqrt{M(x,\lambda)}}\frac{d}{dx} + \frac{i\hbar}{4\sqrt{M(x,\lambda)}}\frac{M'(x,\lambda)}{M^{\frac{3}{2}}(x,\lambda)}\right)\psi(x,t)\,dx \\
&= \int \psi^*(x,t)\left(-\frac{i\hbar}{\sqrt{M(x,\lambda)}}\frac{d}{dx} - \frac{i\hbar}{2}\left(\frac{1}{\sqrt{M(x,\lambda)}}\right)'\right)\psi(x,t)\,dx
\end{aligned} \tag{38}$$

Consequently, according to Eq. (26), the GLMO will be

$$\hat{p}_\lambda = \hat{p}_{eff} = -\frac{i\hbar}{\sqrt{M(x,\lambda)}}\frac{d}{dx} - \frac{i\hbar}{2}\left(\frac{1}{\sqrt{M(x,\lambda)}}\right)' \tag{39}$$

which shows again that the Hermitian GLMO has a unique representation.

4. Applications

In this section, as worked examples, we consider two different position-dependent mass distributions, along with some potential models. In each case, the GLMO and the effective potential are obtained for solving the corresponding position-dependent mass Schrödinger equation.

$$\left(\frac{1}{2m_0}\hat{p}_{eff}^2 + u_{eff}(x)\right)\psi(x) = E\psi(x). \tag{40}$$

4.1 Mass distribution $m(x) = m_0(1 + \lambda x)^{-2}$

In what follows, as a useful application of our proposal for a particular form of $m(x, \lambda)$, the generalized linear momentum operator p_{eff} as well as the associated effective potential $u_{eff}(x)$ given in Eqs. (14) and (16), are obtained for several cases of solvable potentials such as the free particle (null potential), the harmonic oscillator, the double well potential, and the inverse square potential combined with a Coulomb-like potential.

Likewise, the corresponding wave functions and the energy spectra are also obtained. So, let us consider the position-dependent mass

$$m(x, \lambda) = m_0(1 + \lambda x)^{-2}, m_0 = constant \tag{41}$$

from which, according with Eq. (39), leads to the generalized linear momentum operator

$$\hat{p}_{eff} = -i\hbar(1 + \lambda x)\frac{d}{dx} + \frac{i\hbar}{2}\lambda \tag{42}$$

and from the Eq. (16) to the effective potential

$$u_{eff} = V(x) - \frac{\hbar^2\lambda^2}{2m_0}\left(\beta + \frac{3}{4} + 4\alpha(\alpha + \beta + 1)\right) \tag{43}$$

At this point, it is worth mentioning that the operator given Eq. (42) is the Hermitian version of the non-Hermitian operator $p_\lambda = -i\hbar(1 + \lambda x)\frac{d}{dx}$ obtained by Costa Filho et al. [9]. Substituting the expressions $\hat{p}_{eff}$ and u_{eff} in the canonical form of the Schrödinger Eq. (40), one gets the differential equation

$$(1 + \lambda x)^2\frac{d^2}{dx^2}\psi(x) + 2\lambda(1 + \lambda x)\frac{d}{dx}\psi(x) + \frac{2m_0}{\hbar^2}\left(E - V(x) + \frac{\hbar^2\lambda^2}{2m_0}(\beta + 1 + 4\alpha(\alpha + \beta + 1))\right)\psi(x) = 0 \tag{44}$$

Next, we are going to solve the above equation for different potentials $V(x)$ by using the mass distribution given in Eq. (41).

4.2 Null potential

To solve Eq. (44), we use the transformation

$$x = \frac{exp\,(\lambda u) - 1}{\lambda} \tag{45}$$

such that

$$\frac{d}{dx} = exp\,(-\lambda u)\frac{d}{du} \tag{46}$$

$$\frac{d^2}{dx^2} = exp\,(-2\lambda u)\frac{d^2}{du^2} - \lambda exp(-2\lambda u)\frac{d}{du} \tag{47}$$

for which Eq. (44) is rewritten as

$$\frac{d^2}{du^2}\psi + \lambda\frac{d}{du}\psi + \frac{2m_0}{\hbar^2}\left(E_n - V(u) + \frac{\hbar^2\lambda^2}{2m_0}(\beta + 1 + 4\alpha(\alpha + \beta + 1))\right)\psi = 0. \tag{48}$$

Next, using

$$\psi(u) = \varphi(u)\,exp\left(-\frac{\lambda}{2}u\right) \tag{49}$$

one leads to

$$-\frac{\hbar^2}{2m_0}\frac{d^2}{du^2}\varphi(u) + V(u)\varphi(u) = \epsilon\varphi(u) \tag{50}$$

with $\epsilon = E + \frac{\hbar^2\lambda^2}{2m_0}\left(\beta + \frac{3}{4} + 4\alpha(\alpha + \beta + 1)\right)$.

Now, let us consider the simplest case of the free particle (null potential $V(u) = 0$), whose differential equation is given by

$$\varphi''_n(u) + \tilde{k}^2\varphi_n(u) = 0 \tag{51}$$

having eigenfunctions $\varphi_n(u) = \exp\left(\pm i\tilde{k}u\right)$ with $\tilde{k}^2 = k^2 + \lambda^2\left(\beta + \frac{3}{4} + 4\alpha(\alpha + \beta + 1)\right) > 0$ and $k^2 = \frac{2m_0}{\hbar^2}E_n$. Hence, by using the transformations given in Eqs. (45) and (49), the solution $\psi(x)$ for the Schrödinger equation in Eq. (40) or rather the differential equation Eq. (44) is

$$\psi_n(x) = \frac{1}{\sqrt{1 + \lambda x}}\,exp\left(\pm i\sqrt{\frac{k^2}{\lambda^2} + \beta + \frac{3}{4} + 4\alpha(\alpha + \beta + 1)}\,\ln(1 + \lambda x)\right). \tag{52}$$

Considering a free particle in an infinite well of length L, the boundary conditions will be $\psi(0) = \psi(L) = 0$ such that $\sqrt{k^2 + \lambda^2\left(\beta + \frac{3}{4} + 4\alpha(\alpha + \beta + 1)\right)}\,\ln(1 + \lambda L) = n\pi$, $n = 1, 2, 3..$from which, the energy spectrum is

$$E_n = \frac{\hbar^2}{2m_0}\frac{n^2\,\pi^2\,\lambda^2}{ln^2(1 + \lambda L)} - \frac{\hbar^2\,\lambda^2}{8m_0}\left(\beta + \frac{3}{4} + 4\alpha(\alpha + \beta + 1)\right). \tag{53}$$

In this case the values $\beta = -1$ and $\alpha = 0$ lead to

$$E_n = \frac{\hbar^2}{2m_0}\frac{n^2\,\pi^2\,\lambda^2}{ln^2(1 + \lambda L)} + \frac{\hbar^2\,\lambda^2}{8m_0} \tag{54}$$

which matches with the result given in Ref. [19]. Namely, when we use the transformation given in Eq. (45), the length $x = L$ transforms into $u = \tilde{L} = \frac{ln\,(1+\lambda L)}{\lambda}$. So, with $\frac{\hbar^2}{2m_0} = 1$, the above energy spectrum results in

$$E_n = \frac{n^2\,\pi^2\,\lambda^2}{\tilde{L}^2} + \frac{\lambda^2}{4} \tag{55}$$

as given in Ref. [19]. In addition, the above E_n modifies the results obtained in Refs. [9, 10].

4.3 Harmonic oscillator

Now, we are going to consider the case of the harmonic oscillator $V(x) = \frac{1}{2}\,kx^2$ in Eq. (44), such that transformations given in Eqs. (45) and (49) lead to

$$-\frac{\hbar^2}{2m_0}\frac{d^2}{du^2}\varphi_n(u) + \frac{k}{2}\left(\frac{\exp(\lambda u) - 1}{\gamma}\right)^2 \varphi_n(u) = \tilde{E}_n\,\varphi_n(u), \tag{56}$$

with $\tilde{E}_n = E_n + \frac{\hbar^2\lambda^2}{2m_0}(\beta + 3/4 + 4\alpha(\alpha + \beta + 1)$. The above equation has Morse-like solutions given by

$$\varphi_n(u) = (z)^{\nu}\, exp\,(-z/2)\, L_n^{2\nu}(z) \tag{57}$$

where $z = 2bexp(\alpha u)$, $n = b - \nu - 1/2$, $\frac{2m_0}{\hbar^2}\tilde{E}_n = \lambda^2(b^2 - \nu^2)$ and $b^2 = \frac{m_0 k}{\hbar^2\lambda^4}$. The condition for having bound states is $n = 0, 1, 2, 3, \ldots ..(2b - 1/2)$. Hence, by using the above definitions, the energy spectrum is

$$E_n = -\frac{\hbar^2\lambda^2}{2m_0}\left(b - \frac{1}{2} - n\right)^2 - \frac{\hbar^2\lambda^2}{2m_0}\left(\beta + \frac{3}{4} + 4\alpha(\alpha + \beta + 1)\right). \tag{58}$$

As before, the particular case $\beta = -1$ and $\alpha = 0$ leads to

$$E_n = -\frac{\hbar^2\lambda^2}{2m_0}\left(b - \frac{1}{2} - n\right)^2 + \frac{\hbar^2\lambda^2}{2m_0}b^2 + \frac{\hbar^2\,\lambda^2}{8m_0} \tag{59}$$

which reduces to Eq. (50) of Ref. [7] and Eq. (28) of Ref. [20] after identifying the corresponding parameters.

Regarding wave functions, from Eq. (49) and (57) these are

$$\psi_n(x) = (2b(1 + \gamma x))^{\nu - 1/2}\, exp\,(-b(1 + \gamma x)) L_n^{2\nu}(2b(1 + \gamma x)) \tag{60}$$

4.4 Double well potential

In this case, we define the transformation $u = 1 + \lambda x$, such that Eq. (44) writes down as

$$\lambda^2 u^2 \frac{d^2\psi(u)}{du^2} + 2\lambda^2 u \frac{d\psi(u)}{du} + \frac{2m_0}{\hbar^2}\left(E - V(u) + \frac{\hbar^2\lambda^2}{2m_0}(\beta + 1 + 4\alpha(\alpha + \beta + 1))\right)\psi(u) = 0 \tag{61}$$

which, after use $\epsilon = -\left[\frac{2m_0}{\hbar^2\lambda^2}E + \beta + 1 + 4\alpha(\alpha + \beta + 1).\right]$

and propose the double well potential

$$V(u) = \frac{\hbar^2\lambda^2}{2m_0}\left(Au^4 - Bu^2\right), \tag{62}$$

it turns into

$$u^2\frac{d^2\psi(u)}{du^2} + 2u\frac{d\psi(u)}{du} + \left(Bu^2 - Au^4\right)\psi(u) = \epsilon\psi(u). \tag{63}$$

At this point, we use the ansatz

$$\psi(u) = u^a\, exp\,(-bu^2)P(u) \tag{64}$$

where a, b and $P(u)$ will be known later. Hence, Eq. (63) leads to

$$P''(u) + \left(\frac{2(a+1)}{u} - 4bu\right)P'(u) + \left[(4b^2 - A)u^2 + \frac{a(a+1) - \epsilon}{u^2} + B - 2b(2a+3)\right]P(u) = 0 \tag{65}$$

In order to simplify the above equation, we select the parameters a and b as follow $4b^2 = A$ and $a(a+1) = \varepsilon$, such that

$$P''(u) + \left(\frac{2(a+1)}{u} - 4bu\right)P'(u) + [B - 2b(2a+3)]P(u) = 0. \tag{66}$$

With the aim of relating this equation with the associated Laguerre differential equation

$$xL_n^{\delta''}(x) + (\delta + 1 - x)L_n^{\delta'}(x) + +nL_n^{\delta}(x) = 0, \tag{67}$$

we apply the change of variable $x = cz^2$, c being an arbitrary constant, such that

$$\frac{d^2}{dz^2}L_n^{\delta}(cz^2) + \left(\frac{2\delta+1}{z} - 2cz\right)\frac{d}{dz}L_n^{\delta}(cz^2) + 4cnL_n^{\delta}(cz^2) = 0. \tag{68}$$

Comparing Eq. (66) with Eq. (68) leads to

$$2(a+1) = 2\delta + 1, B - 2b(2a+3) = 4cn \text{ and } P(u) = L_n^{\delta}(cz^2) \tag{69}$$

from which, together with the definitions given above $4b^2 = A$ and $a(a+1) = \varepsilon$, leads to $\varepsilon_n = -\left(2n + 1 - \frac{B}{2\sqrt{A}}\right)^2 - \frac{1}{4}$. Hence the energy spectrum results in

$$E_n = -\frac{\hbar^2\lambda^2}{2m_0}\left(2n + 1 - \frac{B}{2\sqrt{A}}\right)^2 - \frac{\hbar^2\lambda^2}{2m_0}\left[\beta + \frac{3}{4} + 4\alpha(\alpha + \beta + 1)\right] \tag{70}$$

and the wave function will be

$$\psi_n(u) = u^{\frac{B}{2\sqrt{A}} - 2n - \frac{3}{2}}e^{-\frac{\sqrt{A}}{2}u^2}L_n^{\frac{B}{2\sqrt{A}} - 2n - 1}\left(\sqrt{A}u^2\right). \tag{71}$$

It is worth noting that the particular case of the ambiguity parameters $\beta = -1$, $\alpha = 0$ leads to

$$E_n = -\frac{\hbar^2\lambda^2}{2m_0}\left(2n+1-\frac{B}{2\sqrt{A}}\right)^2 + \frac{\hbar^2\lambda^2}{8m_0}, \tag{72}$$

which, other than the factor $\frac{\hbar^2\lambda^2}{8m_0}$ and the shifting $n \to n-(1/4)$, is in agreement with Vubangsi et al. [21].

4.5 Inverse square and coulomb-like potential

In this case, we are proposing the potential

$$V(x) = \frac{A}{x^2} - \frac{B}{x} \tag{73}$$

with the transformation $exp\,(-\lambda r) = 1 + \lambda x$ such that Eq. (44) leads to

$$\frac{d^2\psi(r)}{dr^2} - \lambda\frac{d\psi(r)}{dr} + \frac{2m_0}{\hbar^2}\left[\epsilon - \frac{A\lambda^2}{(exp\,(-\lambda r)-1)^2} + \frac{B\lambda}{exp\,(-\lambda r)-1}\right]\psi(r) = 0 \tag{74}$$

where

$$\epsilon = E + \frac{\hbar^2\lambda^2}{2m_0}(\beta+1+4\alpha(\alpha+\beta+1)) \tag{75}$$

So, with the aim of eliminating the first derivative in the above equation, we use the ansatz $\psi(r) = \varphi(r)\,exp\,(\lambda r/2)$ such that

$$-\frac{\hbar^2}{2m_0}\frac{d^2\varphi(r)}{dr^2} + \frac{2m_0}{\hbar^2}\left[\frac{A\lambda^2}{(1-exp\,(-\lambda r))^2} + \frac{B\lambda}{1-exp\,(-\lambda r)}\right]\varphi(r) = \left(\epsilon - \frac{\hbar^2\lambda^2}{8m_0}\right)\varphi(r) \tag{76}$$

At this point, by using the identities

$$\frac{1}{1-exp\,(-\lambda r)} = 1 + \frac{exp\,(-\lambda r)}{1-exp\,(-\lambda r)} \tag{77}$$

$$\frac{1}{(1-exp\,(-\lambda r))^2} = 1 + \frac{exp\,(-\lambda r)}{1-exp\,(-\lambda r)} + \frac{exp\,(-\lambda r)}{(1-exp\,(-\lambda r))^2} \tag{78}$$

the potential in Eq. (76) can be written as an exponential-type potential

$$V(r) = \frac{A\lambda^2}{(1-exp\,(-\lambda r))^2} + \frac{B\lambda}{1-exp\,(-\lambda r)} = (A\lambda^2+B\lambda) + \frac{(A\lambda^2+B\lambda)\;exp\,(-\lambda r)}{1-exp\,(-\lambda r)} + \frac{A\lambda^2\;exp\,(-\lambda r)}{(1-exp\,(-\lambda r))^2} \tag{79}$$

In this context, according to the approach used to solve the multiparameter exponential-type potentials [22, 23], the potential

$$V(r) = \frac{\mathbb{A}\, q\ exp\,(-r/k)}{1 - q exp(-r/k)} + \frac{\mathbb{B}\, q\ exp\,(-r/k)}{(1 - q exp(-r/k))^2} + \frac{\mathbb{C}\, q^2\ exp\,(-2r/k)}{(1 - q exp(-r/k))^2} \tag{80}$$

whose solutions are given in terms of the hypergeometric functions, has energy spectra

$$E_n = -\frac{\hbar^2}{2m_0}\left(\frac{1}{4k}\right)^2\left(2n + 1 + \zeta - \frac{8m_0}{\hbar^2}\frac{k^2(\mathbb{C} - \mathbb{A})}{2n + 1 + \zeta}\right)^2 + \frac{\hbar^2\lambda^2}{8m_0}, \tag{81}$$

with $\zeta = \sqrt{1 + \frac{8m_0k^2}{\hbar^2}(\mathbb{B} + \mathbb{C})}$. So, by comparing the potentials in Eqs. (79) and (80), we can identify the parameters

$\mathbb{A} = A\lambda 2 + B\lambda, = A\lambda 2, \mathbb{C} = 0, q = 1, k = \lambda^{-1}$ and the energy spectrum of the Eq. (76) results in

$$E_n = \epsilon - \frac{\hbar^2\lambda^2}{8m_0} = -\frac{\hbar^2\lambda^2}{8m_0}\left(n + \frac{1}{2} + \frac{1}{2}\zeta + \frac{2m_0}{\hbar^2\lambda^2}\frac{A\lambda^2 + B\lambda}{n + \frac{1}{2} + \frac{1}{2}\zeta}\right)^2 + A\lambda^2 + B\lambda, \tag{82}$$

such that, from Eq. (75)

$$E_n = -\frac{\hbar^2\lambda^2}{8m_0}\left(n + \frac{1}{2} + \frac{1}{2}\zeta + \frac{2m_0}{\hbar^2\lambda^2}\frac{A\lambda^2 + B\lambda}{n + \frac{1}{2} + \frac{1}{2}\zeta}\right)^2 + A\lambda^2 + B\lambda - \frac{\hbar^2\lambda^2}{2m_0}\left(\beta + \frac{3}{4} + 4\alpha(\alpha + \beta + 1)\right) \tag{83}$$

with $\zeta = \sqrt{1 + \frac{8A\, m_0}{\hbar^2}}$

The particular case $\beta = -1$ and $\alpha = 0$ leads to

$$E_n = -\frac{\hbar^2\lambda^2}{8m_0}\left(n + \frac{1}{2} + \frac{1}{2}\zeta + \frac{2m_0}{\hbar^2\lambda^2}\frac{A\lambda^2 + B\lambda}{n + \frac{1}{2} + \frac{1}{2}\zeta}\right)^2 + \frac{\hbar^2\lambda^2}{8m_0} + A\lambda^2 + B\lambda, \tag{84}$$

which, other than the factor $\frac{\hbar^2\lambda^2}{8m_0} + A\lambda^2 + B\lambda$, matches with Eq. (16) of Ref. [24].

Likewise, according to the multiparameter exponential-type potential method mentioned before, the wave function is given in terms of the hypergeometric function as

$$\psi_n(u) = (exp\,(-\lambda r))^{\frac{b-c-n}{2}}\,(1 - exp\,(-\lambda r))^{\frac{\zeta}{2}}\,{}_2F_1(a, b, c; 1 - exp\,(-\lambda r)) \tag{85}$$

where the hypergeometric parameters are $a = -n$, $b = -\frac{2m_0}{\hbar^2\lambda^2}\frac{A\lambda^2 + B\lambda}{n + \frac{1}{2} + \frac{1}{2}\zeta}$, $c = 1 + \zeta$.

4.6 Mass distribution $m(x) = m_0(1 + \lambda^2x^2)^{-1}$

In order to consider the position-dependent mass distribution already presented by da Costa et al. [8], in this case, for $m(x, \lambda) = m_0(1 + \lambda^2x^2)^{-1}$, by using the Eqs. (39) and (16), the corresponding GLMO and the effective potential are

$$\hat{p}_{eff} = -i\hbar\sqrt{1+\lambda^2x^2}\frac{d}{dx} - \frac{i\hbar\lambda^2 x}{2\sqrt{1+\lambda^2x^2}} \tag{86}$$

and

$$u_{eff}(x) = V(x) - \frac{\hbar^2\lambda^2}{2m_0}\left(4\alpha(\alpha+\beta+1)+\frac{1}{4}\right)\frac{\lambda^2x^2}{1+\lambda^2x^2} + \frac{\hbar^2\lambda^2}{2m_0}\left(\beta+\frac{1}{2}\right), \tag{87}$$

while the canonical PDM Schrödinger Eq. (40) is written down as

$$\begin{aligned}&(1+\lambda x)\frac{d^2\psi(x)}{dx^2} + 2\lambda^2 x\frac{d\psi(x)}{dx} + \frac{2m_0}{\hbar^2}\left(E - V(x) + \frac{\hbar^2\lambda^2}{2m_0}\left(\beta+\frac{1}{2}\right) + \frac{\hbar^2\lambda^2}{2m_0}4\alpha(\alpha+\beta++1)\frac{\lambda^2x^2}{1+\lambda^2x^2}\right)\\ &\psi(x) = 0\end{aligned} \tag{88}$$

Then, by using the transformation

$$x = \frac{1}{\lambda}\,sinh\,(\lambda u) \tag{89}$$

the above equation results in

$$\begin{aligned}&\frac{d^2\psi(u)}{du^2} + \lambda\,tanh\,(\lambda u)\frac{d\psi(u)}{du} + \frac{2m_0}{\hbar^2}\left(E - V(u) + \frac{\hbar^2\lambda^2}{2m_0}\left(\beta+\frac{1}{2}\right) + \frac{\hbar^2\lambda^2}{2m_0}4\alpha(\alpha+\beta+1)\,tanh^2(\lambda u)\right)\\ &\psi(u) = 0,\end{aligned} \tag{90}$$

$$\begin{aligned}&\frac{d^2\psi(u)}{du^2} + \lambda\,tanh\,(\lambda u)\frac{d\psi(u)}{du} + \frac{2m_0}{\hbar^2}\left(E - V(u) + \frac{\frac{1}{2}\hbar^2\lambda^2}{2m_0}(2\beta+1) + \frac{\hbar^2\lambda^2}{2m_0}4\alpha(\alpha+\beta+1)\,tanh^2(\lambda u)\right)\\ &\psi(u) = 0\end{aligned} \tag{91}$$

which, after applying the similarity condition

$$\psi(u) = \varphi(u)\,exp\left(-\frac{\lambda}{2}\int tanh\,(\lambda u)du\right) = \varphi(u)\,exp\left(-\frac{1}{2}\,ln\,(cosh\,(\lambda u))\right) \tag{92}$$

becomes

$$\begin{aligned}&\frac{d^2\varphi(u)}{du^2} + \frac{2m_0}{\hbar^2}\left(E - V(u) + \frac{\frac{1}{4}\hbar^2\lambda^2}{2m_0}(4\beta+3) + \frac{\hbar^2\lambda^2}{2m_0}\left(4\alpha(\alpha+\beta+1)\,tanh^2(\lambda u) - \frac{1}{4}sech^2(\lambda u)\right)\right)\\ &\varphi(u) = 0\end{aligned} \tag{93}$$

Also, to further simplify this equation, we propose as before the parameters $\alpha = 0$, $\beta = -1$ to get

$$-\frac{\hbar^2}{2m_0}\frac{d^2\varphi(u)}{du^2} + \left(V(u) + \frac{\hbar^2\lambda^2}{8m_0}sech^2(\lambda u)\right)\varphi(u) = \left(E - \frac{\hbar^2\lambda^2}{8m_0}\right)\varphi(u) \tag{94}$$

A simple case would be when the potential is $V(u) = \frac{\hbar^2\lambda^2}{8m_0} tanh^2(\lambda u)$, such that one has

$$\frac{d^2\varphi(u)}{du^2} + k^2\varphi(u) = 0 \tag{95}$$

with $k^2 = \frac{\hbar^2\lambda^2}{2m_0}E - \frac{\lambda^2}{2}$. Then $\varphi(u) = exp\,(\pm i\lambda u) = exp\left(\pm iln\left(\lambda x + \sqrt{1 + \lambda^2x^2}\right)\right)$ and according with Eq. (92)

$$\psi(x) = exp\left(-\frac{1}{2}\,ln\left(cosh\left(\lambda x + \sqrt{1 + \lambda^2x^2}\right)\right)\right)exp\left(\pm iln\left(\lambda x + \sqrt{1 + \lambda^2x^2}\right)\right) \tag{96}$$

Regarding with the λ-deformed coordinate and linear momentum operators are respectively

$$x_\lambda = \int\sqrt{M(x,\lambda)}\,dx = \frac{1}{\lambda}\,ln\left(\lambda x + \sqrt{1 + \lambda^2x^2}\right) \tag{97}$$

and

$$p_\lambda = -i\hbar\sqrt{1 + \lambda^2x^2}\frac{d}{dx} - \frac{i\hbar\lambda^2x}{2\sqrt{1 + \lambda^2x^2}} \tag{98}$$

The above equations correspond to those given in Eqs. [31a,b] of Ref. [8] with properties

$$[x_\lambda, p_\lambda] = i\hbar \tag{99}$$

such that $\lambda \to 0$ leads to standard variables $x_\lambda \to x,\ p_\lambda \to -i\hbar\frac{d}{dx}$.

In the particular case of the mass under study, the corresponding λ-deformed exponential function comes from the Eq. (33) as

$$exp_\lambda(x) = exp\left(\frac{1}{\lambda}\,ln\left(\lambda x + \sqrt{1 + \lambda^2x^2}\right)\right) = \left(\lambda x + \sqrt{1 + \lambda^2x^2}\right)^{\frac{1}{\lambda}}, \tag{100}$$

while the λ-deformed logarithmic function will be

$$ln_\lambda(x) = \frac{1}{\lambda}\,sinh\,(\lambda ln(x)) = \frac{x^\lambda - x^{-\lambda}}{2\lambda} \tag{101}$$

in agreement with Eqs. (1) and (2) proposed recently by da Costa et al. [8].

5. Conclusions

The purpose of this work has been twofold: firstly, to generalize the quantum linear momentum operator (GLMO), and secondly to show that the GLMO is unique independently of the ambiguity parameters. To attain the first objective, we have considered the canonical form of the position-dependent mass Hamiltonian, which means to write the von-Ross's Hamiltonian in terms of the Hermitian GLMO plus a potential energy function, assuming arbitrary ambiguity parameters. To achieve the second purpose, in addition to the method used to fulfill the first objective, we have used three other different approaches: namely, factorization of the von-Roos Hamiltonian along with a hermiticity condition, by using an operator that leaves invariant a λ-deformed exponential function, and by analogy to the procedure used to obtain the standard linear momentum operator. Our proposal is general for any position-dependent mass distribution $m(x, \lambda)$ and ambiguity parameters. In fact, as a useful application of the proposed method, as examples, we have considered two particular $m(x, \lambda)$, as well as some different potential models: null potential, the harmonic oscillator, the double-well potential, and the Coulomb-like potential. Our results are generalizations of particular findings already published. That is, for some specific values of the ambiguity parameters, we are in agreement with known results, which means that our method improves those studies on the same subject given until now.

Acknowledgements

This work was partially supported by the projects UAM-A-CBI-2232004 and 009. One of us (JGR) thanks to the National Polytechnic Institute Mexico, for the financial support given through the COFAA-IPN project SIP-20240064. We are grateful to the SNII - Conahcyt - México for the stipend received.

Author details

Jesús García-Ravelo[1], Jesús García-Martínez[2], Jesús Morales[3] and José Juan Peña[3]*

1 National Polytechnic Institute, ESFM, CDMX, México

2 Biomedical Engineering Division, Ixtapaluca Institute of Advanced Studies, Estado de México

3 Metropolitan Autonomous University, Azc. CDMX, México

*Address all correspondence to: jjpg@azc.uam.mx

References

[1] Barsan V, Ciornei MC. Semiconductor quantum wells with BenDaniel-Duke boundary conditions: Approximate analytical results. European Journal of Physics. 2017;**38**:015407. DOI: 10.1088/0143-0807/38/1/015407

[2] Chang LN, Minic D, Okamura N. Exact solution of the harmonic oscillator in arbitrary dimensions with minimal length uncertainty relations. Physical Review D. 2002;**65**:125027. DOI: 10.1103/PhysRevD.65.125027

[3] Vubangsi M, Tchoffo M, Fai LC. New kinetic energy operator for variable mass systems. European Physical Journal Plus. 2014;**129**:105. DOI: 10.1140/epjp/i2014-14105-4

[4] Tsallis C. Possible generalization of Boltzmann- Gibbs statistics. Journal of Statistical Physics. 1988;**52**:479. DOI: 10.1007/BF01016429

[5] Tsallis C. Introduction to Nonextensive Statistical Mechanics: Approaching a Complex World. New York: Springer; 2009. ISBN 978-0-387-85359-8

[6] da Costa BG, Borges EP. Generalized space and linear momentum operators in quantum mechanics. Journal of Mathematical Physics. 2014;**55**:062105. DOI: 10.1063/1.4884299

[7] da Costa BG, Borges EP. A position-dependent mass harmonic oscillator and deformed space. Journal of Mathematical Physics. 2018;**59**:042101. DOI: 10.1063/1.5020225

[8] da Costa BG, Gomez IS, Portesi M. κ-Deformed quantum and classical mechanics for a system with position-dependent effective mass. Journal of Mathematical Physics. 2020;**61**:082105. DOI: 10.1063/5.0014553

[9] Costa Filho RN, Almeida MP, Farias GA, Andrade JS. Displacement operator for quantum systems with position-dependent mass. Physical Review A. 2011;**84**:050102(R). DOI: 10.1103/PhysRevA.84.050102

[10] Mazharimousavi SH. Revisiting the displacement operator for quantum systems with position-dependent mass. Physical Review A. 2012;**85**:034102. DOI: 10.1103/PhysRevA.85.034102. Erratum: Phys. Rev. A 89, 049904 (2014) DOI: 10.1103/PhysRevA.89.049904

[11] Quesne C, Tkachuk VM. Deformed algebras, position-dependent effective masses and curved spaces: An exactly solvable Coulomb problem. Journal of Physics A: Mathematical and General. 2004;**37**:4267. DOI: 10.1088/0305-4470/37/14/006

[12] Kempf A, Mangano G, Mann RB. Hilbert space representation of the minimal length uncertainty relation. Physical Review D. 1995;**52**:1108. DOI: 10.1103/PhysRevD.52.1108

[13] von Roos O. Position-dependent effective masses in semiconductor theory. Physical Review B. 1983;**27**:7547. DOI: 10.1103/PhysRevB.27.7547

[14] Li TL, Kuhn KJ. Band-offset ratio dependence on the effective-mass Hamiltonian based on a modified profile of the $GaAs_x$-$AlGa_{1-x}$ As quantum well. Physical Review B.1993;**47**:12760. DOI: 10.1103/PhysRevB.47.12760

[15] Zhu QG, Kroemer H. Interface connection rules for effective-mass wave functions at an abrupt heterojunction between two different semiconductors.

Physical Review B. 1983;**27**:3519. DOI: 10.1103/PhysRevB.27.3519

[16] Daniel DJB, Duke CB. Space-charge effects on electron tunneling. Physical Review Journals Archive. 1966;**152**:683. DOI: 10.1103/PhysRev.152.683

[17] Rego-Monteiro A, Rodrigues LMCS, Curado EMF. Position dependent mass quantum Hamiltonians: General approach and duality. Journal of Physics A: Mathematical and Theoretical. 2016; **49**:125203. DOI: 10.1088/1751-8113/49/12/125203

[18] Lèvy-Leblon JM. Elementary quantum models with position-dependent mass. European Journal of Physics. 1992;**13**:215. DOI: 10.1088/0143-0807/13/5/003

[19] Peña JJ, Ovando G, Morales J, García-Ravelo J, Pacheco-García C. Exactly solvable Schrödinger equations with a position-dependent mass: Null potential. International Journal of Quantum Chemistry. 2007;**107**:3039. DOI: 10.1002/qua.21526

[20] Pacheco-García C, García-Ravelo J, Morales J, Peña JJ. Exactly solvable effective mass Schrödinger equation with coulomb-like potential. International Journal of Quantum Chemistry. 2010;**110**:2880. DOI: 10.1002/qua.22898

[21] Vubangsi M, Tchoffo M, Fai LC. Position-dependent effective mass system in a variable potential: Displacement operator method. Physica Scripta. 2014;**89**:025101. DOI: 10.1088/0031-8949/89/02/025101

[22] Peña JJ, Ovando G, Morales J, García-Ravelo J. Non-deformed singular and non-singular exponential-type potentials. Journal of Molecular Modeling. 2017;**23**:265. DOI: 10.1007/s00894-017-3423-8

[23] Peña J, J, García-Ravelo J, Ovando G, Morales J. Approximate ℓ-bound state solutions of q-deformed exponential-type potentials. International Journal of Quantum Chemistry. 2020;**120**:111, e26189. DOI: 10.1002/qua.26189

[24] Arda A, Sever R. Effective mass quantum systems with displacement operator: inverse square plus Coulomb-like potential. Few-Body Systems. 2015; **56**:697. DOI: 10.1007/s00601-015-1008-6

Chapter 5

Practical Stabilization of Nonlinear Cascade Systems and Applications

Ines Ellouze and Maryam Ben Salah

Abstract

In this chapter, we are interested in the problem of the global stabilization of certain nonlinear cascaded systems. Furthermore, we study the global exponential stabil- ity of nonlinear non-autonomous systems. Then, we investigate the global practi- cal stabilization of non-autonomous cascaded systems which we give an illustrative example. Otherwise, we present the practical stabilization of perturbed cascaded non-autonomous systems and the global Practical stabilization by output closed loop.

Keywords: cascaded system, practical stabilization, non-autonomous system, nonlinear system, exponential stability

1. Introduction

In this chapter, we are interested in the problem of the global stabilization of certain nonlinear systems, the so-called cascade systems of the following form:

$$\begin{cases} \dot{x}_1 = F(t, x_1, x_2, u) \\ \dot{x}_2 = G(t, x_2). \end{cases} \tag{1}$$

where $x = (x_1, x_2) \in \mathbb{R}^n$ and $u \in \mathbb{R}^p$. The functions F and G are continuous at t and locally Lipschitz at x. For this class of system, it is well known that the local asymptotic stability of each subsystem, namely $\dot{x}_1 = F(t, x_1, 0, u)$ and $\dot{x}_2 = G(t, x_2)$, results in the local asymptotic stability of the compound system [1]. This local result has no global analog. In fact, the global stability can be deduced from the subsystems only with additional conditions. Seibert and Suarez [2] showed, in the autonomous case, that if each subsystem is globally asymptotically stable and all the orbits of the compound system are bounded, then the compound system remains globally asymptotically stable.

Another approach is to use Lyapunov techniques. Assuming that each subsystem is globally uniformly asymptotically stable, a Lyapunov function of the compound system can be constructed, allowing one to deduce the result [3–6].

The stabilization of nonlinear systems has been widely studied by several authors (see Refs. [2, 6–9]).

In this chapter, we try to determine stabilizing control laws, under sufficient conditions, for the nonlinear systems in cascade (1).

2. Global exponential stability of nonlinear nonautonomous systems

We consider the nonlinear system of the form:

$$\dot{x} = f(t,x) + g(t,x), \tag{2}$$

where $f, g : \mathbb{R}_+ \times \mathbb{R}^n \to \mathbb{R}^n$ are continuous and locally Lipschitz at x.
We introduce the following hypotheses:

- $(\mathcal{A}_1)$ $x = 0$ is a globally uniformly exponentially stable equilibrium point of the system

$$\dot{x} = f(t,x)$$

with the Lyapunov function, $V : \mathbb{R}_+ \times \mathbb{R}^n \to \mathbb{R}_+$ which satisfies

$$c_1\|x\|^2 \le V(t,x) \le c_2\|x\|^2$$
$$\frac{\partial V}{\partial t}(t,x) + \frac{\partial V}{\partial x}(t,x)f(t,x) \le -c_3V(t,x)$$
$$\left\|\frac{\partial V}{\partial x}\right\| \le c_4\|x\|$$

where c_1, c_2, c_3, c_4 are strictly positive constants.

- $(\mathcal{A}_2)$ There exists a continuous positive function $\gamma : \mathbb{R}_+ \to \mathbb{R}_+$, verifies

$$\|g(t,x)\| \le \gamma(t)\|x\|, \quad \forall x \in \mathbb{R}^n$$

with

$$\gamma(t) < \frac{c_1c_3}{c_4},$$
$$\int_0^{+\infty} \gamma(s)\,ds \le M_\gamma < +\infty.$$

where M_γ is a positive constant.

Theorem 1.2.1. *Under the assumptions $(\mathcal{A}_1 - \mathcal{A}_2)$ the system* (2) *is globally uniformly exponentially stable.*

Proof:

The derivative of the function V along the trajectories of the system (2) is given by:

$$\dot{V}(t,x) = \frac{\partial V}{\partial t}(t,x) + \frac{\partial V}{\partial x}(t,x)(f(t,x) + g(t,x)).$$

Taking into account $(\mathcal{A}_1)$ and $(\mathcal{A}_2)$, we will have

$$\begin{aligned}\dot{V}(t,x) &\leq -c_3V(t,x) + \left\|\frac{\partial V}{\partial x}\right\| \|g,(t,x)\| \\ &\leq -c_3V(t,x) + c_4\gamma(t)\|x\|^2 \\ &\leq -\left(c_3 - \frac{c_4}{c_1}\gamma(t)\right)V(t,x).\end{aligned}$$

Integrating between t_0 and t, we obtain

$$V(t,x) \leq V(t_0,x_0)e^{-\left(\frac{c_3c_1-c_4M_\gamma}{c_1}\right)(t-t_0)}, \quad \forall t > t_0 \geq 0.$$

It follows that

$$\|x,(t)\| \leq \sqrt{\frac{c_2}{c_1}}\|x_0\|e^{-\left(\frac{c_3c_1-c_4M_\gamma}{2c_1}\right)(t-t_0)}.$$

Then, the system (2) is globally uniformly exponentially stable.

3. Global practical stabilization of nonautonomous cascaded systems

This part is devoted to the problem of stabilization by state feedback of systems of the form:

$$\begin{cases} \dot{x}_1 = f_1(t,x_1,x_2) + g_1(t,x_1)[u + P(t,x_1,x_2)K(t,x_1,x_2)] \\ \dot{x}_2 = f_2(t,x_2) + g_2(t,x_2), \end{cases} \tag{3}$$

where $x = (x_1,x_2) \in \mathbb{R}^p \times \mathbb{R}^q$, $u \in \mathbb{R}^p$ respectively denote the state, control (input) of the system, $f_1 : \mathbb{R}_+ \times \mathbb{R}^p \times \mathbb{R}^q \to \mathbb{R}^p$, $g_1 : \mathbb{R}_+ \times \mathbb{R}^p \to \mathbb{R}$ and $f_2, g_2 : \mathbb{R}_+ \times \mathbb{R}^q \to \mathbb{R}^q$ are continuous and locally Lipschitz at x.

$P : \mathbb{R}_+ \times \mathbb{R}^p \times \mathbb{R}^q \to \mathbb{R}$ is a continuous function and $K : \mathbb{R}_+ \times \mathbb{R}^p \times \mathbb{R}^q \to \mathbb{R}^p$ is the unknown function satisfying the following property:

$(\mathcal{P})$ There exists a positive real function $\rho(t,x_2)$ such that

$$\|K(t,x_1,x_2)\| \leq \rho(t,x_2)\|x_1\|. \tag{4}$$

Our goal is to construct a controller such that the system (3) is globally uniformly practically exponentially stable. We then introduce the following hypotheses.

$(\mathcal{H}_1)$ There is a feedback $\alpha(t,x_1)$, a function $W : \mathbb{R}_+ \times \mathbb{R}^p \to \mathbb{R}$ continuously differentiable verifying

$$\begin{gathered} a_1\|x_1\|^2 \leq W(t,x_1) \leq a_2\|x_1\|^2 \\ \frac{\partial W}{\partial t}(t,x_1) + \frac{\partial W}{\partial x_1}(t,x_1)\left[f_1(t,x_1,0) + g_1(t,x_1)\alpha(t,x_1)\right] \leq -a_3\|x_1|^2 \\ \left\|\frac{\partial W}{\partial x_1}\right\| \leq a_4\|x_1\| \end{gathered} \tag{5}$$

where a_1, a_2, a_3, a_4 are strictly positive constants with $a_2 < 4a_1a_3$

$(\mathcal{H}_2)$ There exists a continuous positive function $\lambda : \mathbb{R}_+ \to \mathbb{R}_+$ such that

$$\|f_1(t,x_1,x_2) - f_1(t,x_1,0)\| \leq \lambda(t)\|x_2\|, \quad \forall\ t > t_0 \geq 0, \quad \forall (x_1,x_2) \in \mathbb{R}^p \times \mathbb{R}^q$$

with

$$\int_0^{+\infty} \lambda(s)\,ds \leq M_\lambda < +\infty \quad et \quad \int_0^{+\infty} (\lambda(s))^2 ds \leq M'_\lambda < +\infty,$$

where M_λ, M'_λ are positive constants.

$(\mathcal{H}_3)$ There exist two constants $k, \gamma > 0$, such that the solutions of the subsystem

$$\dot{x}_2 = f_2(t,x_2) + g_2(t,x_2),$$

verify:

$$\|x_2(t)\| \leq k\|x_{2_0}\|e^{-\gamma(t-t_0)}, \quad \forall x_{2_0} \in \mathbb{R}^q, \quad \forall t > t_0 \geq 0.$$

Theorem 1.3.1. *Consider the system* (3) *verifying the hypotheses* $(\mathcal{H}_1)$*-*$(\mathcal{H}_3)$ *with the nonlinear feedback*

$$u(t,x_1,x_2) = \alpha(t,x_1) - \left(\frac{\partial W}{\partial x_1} g_1(t,x_1)\right)^T \|P,(t,x_1,x_2)\|^2 \rho^2(t,x_2). \tag{6}$$

Then the resulting closed-loop system is globally, uniformly, practically exponentially stable.

Proof:

We will consider the function W as a Lyapunov function for the looped system (3-6). The derivative of W along the trajectories of (3-6) is given by:

$$\dot{W}(t,x_1) = \frac{\partial W}{\partial t} + \frac{\partial W}{\partial x_1}(f_1(t,x_1,x_2) + g_1(t,x_1)[u(t,x_1,x_2) + P(t,x_1,x_2)K(t,x_1,x_2)],)$$

Using $(\mathcal{H}_1)$, we obtain

$$\begin{aligned}
\dot{W}(t,x_1) &\leq \frac{\partial W}{\partial t} + \frac{\partial W}{\partial x_1}(f_1(t,x_1,0) + g_1(t,x_1)\alpha(t,x_1),) + \frac{\partial W}{\partial x_1}(f_1(t,x_1,x_2) - f_1(t,x_1,0),) \\
&- \left\|\frac{\partial W}{\partial x_1} g_1(t,x_1)\right\|^2 \|P(t,x_1,x_2)\|^2 \rho^2(t,x_2) + \frac{\partial W}{\partial x_1} g_1(t,x_1)P(t,x_1,x_2)K(t,x_1,x_2) \\
&\leq -a_3\|x_1\|^2 + \left\|\frac{\partial W}{\partial x_1}\right\| \|f_1(t,x_1,x_2) - f_1(t,x_1,0)\| \\
&- \left\|\frac{\partial W}{\partial x_1} g_1(t,x_1)\right\|^2 \|P(t,x_1,x_2)\|^2 \rho^2(t,x_2) \\
&+ \left\|\frac{\partial W}{\partial x_1} g_1(t,x_1)\right\| \|P(t,x_1,x_2)\| \|K(t,x_1,x_2)\|.
\end{aligned}$$

By virtue of $(\mathcal{H}_2)$ and the property $(\mathcal{P})$, we will have

$$\begin{aligned}\dot{W}(t,x_1) &\le -a_3\|x_1\|^2 + a_4\|x_1\|(\lambda(t)\|x_2\|)\\ &- \left[\frac{1}{2}\|x_1\| - \left\|\frac{\partial V_1}{\partial x_1} g_1(t,x_1)\right\| \|P(t,x_1,x_2)\|\rho(t,x_2)\right]^2 + \frac{1}{4}\|x_1\|^2.\end{aligned}$$

So, for all $\varepsilon > 0$

$$\dot{W}(t,x_1) \le -a_3\|x_1\|^2 + \frac{1}{4}\|x_1\|^2 + a_4\lambda_1(t)\left(\frac{1}{2\varepsilon}\|x_1\|^2 + \frac{\varepsilon}{2}\|x_2\|^2\right).$$

Which implies that, using the conditions (5) and $(\mathcal{H}_3)$,

$$\begin{aligned}\dot{W}(t,x_1) &\le -\left(\frac{a_3}{a_2} - \frac{a_4}{2a_1\varepsilon}\lambda(t) - \frac{1}{4a_1}\right)W(t,x_1) + \frac{\varepsilon a_4\lambda(t)}{2}\|x_2\|^2.\\ &\le -\left(\frac{a_3}{a_2} - \frac{a_4}{2a_1\varepsilon}\lambda(t) - \frac{1}{4a_1}\right)W(t,x_1) + \frac{\varepsilon a_4\lambda(t)}{2}\left(k\|x_{2_0}\|e^{-\gamma(t-t_0)}\right)^2.\end{aligned}$$

Let us choose ε such that,

$$\lambda(t) < \left(\frac{2a_1a_3}{a_2a_4} - \frac{1}{2a_4}\right)\varepsilon, \quad \forall t \ge 0.$$

To simplify the previous presentation, let us put $\alpha(t) = \frac{a_3}{a_2} - \frac{a_4}{2a_1\varepsilon}\lambda(t) - \frac{1}{4a_1} > 0$, and $\kappa(t) = \frac{\varepsilon a_4\lambda(t)}{2}\left(k\|x_{2_0}\|e^{-\gamma(t-t_0)}\right)^2$.
And subsequently

$$\dot{W}(t,x_1) \le -\alpha(t)W(t,x_1) + \kappa(t).$$

Let $z(t) = W(t,x_1)\, e^{\int_{t_0}^t \alpha(s)\,ds}, \quad \forall\, t \ge t_0 \ge 0.$
It follows that

$$\begin{aligned}\dot{z}(t) &= \left(\dot{W}(t,x_1) + \alpha(t)W(t,x_1)\right)e^{\int_{t_0}^t \alpha(s)\,ds}\\ &\le \kappa(t)e^{\int_{t_0}^t \alpha(s)\,ds}\end{aligned}$$

Integrating between t_0 and t, we obtain $\forall \quad t \ge t_0$,

$$z(t) \le z(t_0) + \int_{t_0}^t \kappa(s)e^{\int_{t_0}^s \alpha(\tau)\,d\tau}\,ds.$$

And subsequently

$$\begin{aligned}W(t,x_1) &\le W(t_0,x_{1_0})e^{-\int_{t_0}^t \alpha(s)\,ds}\\ &+ \left[\int_{t_0}^t \kappa(s)e^{\int_{t_0}^s \alpha(\tau)\,d\tau}\,ds\right]e^{-\int_{t_0}^t \alpha(s)\,ds}.\end{aligned}$$

Or

$$\int_{t_0}^{t} \alpha(s)ds = \frac{4a_1a_3 - a_2}{4a_1a_2}(t - t_0) - \frac{a_4}{2a_1\varepsilon}\int_{t_0}^{t} \lambda(s)ds,$$

Then

$$e^{\int_{t_0}^{t} \alpha(s)\,ds} \leq e^{\frac{4a_1a_3-a_2}{4a_1a_2}(t-t_0)}, \tag{7}$$

which implies

$$e^{-\int_{t_0}^{t} \alpha(s)\,ds} = e^{\frac{a_4M_\lambda}{2a_1\varepsilon}}\, e^{-\frac{4a_1a_3-a_2}{4a_1a_2}(t-t_0)}. \tag{8}$$

According to (7), we will have

$$\begin{aligned}
\int_{t_0}^{t} \kappa(s)e^{\int_{t_0}^{s} \alpha(\tau)\,d\tau} ds \quad &\leq \quad \frac{a_4\varepsilon}{2}(k\|x_{2_0}\|)^2 \left(\int_{t_0}^{t} (\lambda(s))^2 ds\right)^{\frac{1}{2}} \\
&\times \left(\int_{t_0}^{t} \left(e^{-2\gamma(s-t_0)}\, e^{\int_{t_0}^{s} \alpha(\tau)\,d\tau}\right)^2 ds\right)^{\frac{1}{2}} \\
&\leq \quad \frac{a_4\varepsilon}{2}(k\|x_{2_0}\|)^2 \sqrt{M'_\lambda} \\
&\times \left(\left(\int_{t_0}^{t}, e^{-8\gamma(s-t_0)}, ds,\right)^{\frac{1}{2}} \left(\int_{t_0}^{t} e^{4\int_{t_0}^{s} \alpha(\tau)\,d\tau} ds\right)^{\frac{1}{2}}\right)^{\frac{1}{2}} \\
&\leq \quad \frac{a_4\varepsilon}{2}(k\|x_{2_0}\|)^2 \sqrt{M'_\lambda}\left(\frac{a_1a_2}{8\gamma(4a_1a_3 - a_2)}\right)^{\frac{1}{4}} e^{\frac{4a_1a_3-a_2}{4a_1a_2}(t-t_0)}.
\end{aligned}$$

and then, by virtue of (8), we obtain

$$\left(\int_{t_0}^{t} \kappa(s)e^{\int_{t_0}^{s} \alpha(\tau)\,d\tau} ds\right) e^{-\int_{t_0}^{t} \alpha(s)\,ds} \quad \leq \quad \frac{a_4\varepsilon}{2}(k\|x_{2_0}\|)^2 \sqrt{M'_\lambda}\left(\frac{a_1a_2}{8\gamma(4a_1a_3 - a_2)}\right)^{\frac{1}{4}} e^{\frac{a_4M_\lambda}{2a_1\varepsilon}}.$$

It follows that

$$\begin{aligned}
W(t, x_1) \quad &\leq \quad W(t_0, x_{1_0})e^{\frac{a_4M_\lambda}{2a_1\varepsilon}} e^{-\frac{4a_1a_3-a_2}{4a_1a_2}(t-t_0)} \\
&+ \quad \frac{a_4\varepsilon}{2} \;(k\|x_{2_0}\|)^2 \;\sqrt{M'_\lambda}\; \left(\frac{a_1a_2}{8\gamma(4a_1a_3 - a_2)}\right)^{\frac{1}{4}} e^{\frac{a_4M_\lambda}{2a_1\varepsilon}}.
\end{aligned}$$

Using the condition (5), we obtain

$$\begin{aligned}\|x_1(t)\| &\leq \sqrt{\frac{a_2}{a_1}}\ \|x_{1_0}\|\ e^{\frac{a_4 M_\lambda}{4a_1\varepsilon}}\ e^{-\frac{4a_1a_3-a_2}{8a_1a_2}(t-t_0)} \\ &+ \sqrt{\frac{a_4\varepsilon\sqrt{M'_\lambda}}{2a_1}}\ K\ \|x_{2_0}\| \left(\frac{a_1a_2}{8\gamma(4a_1a_3-a_2)}\right)^{\frac{1}{8}} e^{\frac{a_4 M_\lambda}{4a_1\varepsilon}}.\end{aligned}$$

As a result, the closed-loop system (3-6) is globally uniformly practically exponentially stable. □

4. Numerical example

We now give an example of the system of the form (3) that satisfies the assumptions $(\mathcal{H}_1)-(\mathcal{H}_3)$.

Consider the following planar system

$$\begin{cases}\dot{x}_1 = -x_1^3 + \frac{x_2}{1+x_2^2}(1+t)e^{-t} + \left(u + \frac{e^{-\frac{x_2^2}{2}}}{1+t^2}x_1\right), \\ \dot{x}_2 = -3x_2 + x_2 e^{-2t},\end{cases} \tag{9}$$

This system is of the form (3) with

$$f(t,x) = \begin{bmatrix} f_1(t,x_1,x_2) \\ f_2(t,x_2)\end{bmatrix} = \begin{bmatrix} -x_1^3 + \frac{x_2}{1+x_2^2}(1+t)e^{-t} \\ -3x_2 \end{bmatrix},$$

$$g(t,x) = \begin{bmatrix} g_1(t,x_1) \\ g_2(t,x_2)\end{bmatrix} = \begin{bmatrix} 1 \\ x_2 e^{-2t}\end{bmatrix},$$

$P(t,x_1,x_2) = \frac{1}{1+t^2}$ and $K(t,x_1,x_2) = e^{-\frac{x_2^2}{2}}x_1$ which verifies the property $(\mathcal{P})$ with $\rho(t,x_2) = e^{-\frac{x_2^2}{2}}$. Let us take the Lyapunov function $W(t,x_1) = x_1^2$ and $\alpha(t,x_1) = x_1^3 - 3x_1$ then the hypothesis $(\mathcal{H}_1)$ is satisfied with

$$a_1 = a_2 = 1, \alpha_3 = 6, \alpha_4 = 2.$$

Then $(\mathcal{H}_2)$ is verified with $\lambda(t) = (1+t)e^{-t}$, where

$$\int_0^{+\infty}(1+s)e^{-s}ds = 2 < +\infty \quad et \quad \int_0^{+\infty}(1+s)^2 e^{-2s}ds = \frac{5}{4} < +\infty.$$

On the other hand, let us choose the Lyapunov function: $V(t,x_2) = x_2^2$, thus the hypothesis $(\mathcal{H}_3)$ is satisfied with any solution of the subsystem verified:

$$|x_2(t)| \leq |x_{2_0}|e^{-\frac{5}{2}(t-t_0)}$$

Therefore, by applying Theorem 1.3.1, the system (9) in closed-loop by the control

$$u(t,x_1,x_2) = x_1^3 - 3x_1\left(1 + \frac{2e^{-x_2^2}}{3(1+t^2)^2}\right)$$

is globally uniformly exponentially practically stable, and hence the closed system is given by:

$$\begin{cases} \dot{x}_1 = -3x_1\left(1 + \dfrac{2e^{-x_2^2}}{3(1+t^2)^2} - \dfrac{e^{-\frac{x_2^2}{2}}}{3(1+t^2)}\right) + \dfrac{x_2}{1+x_2^2}(1+t)e^{-t} \\ \dot{x}_2 = -3x_2 + x_2 e^{-2t}. \end{cases} \tag{10}$$

We will take as initial condition $x_0 = [2,1]^T$. The result of the simulation is given in Figure.

5. Practical stabilization of perturbed cascaded nonautonomous systems

Now, we consider the perturbed system

$$\begin{cases} \dot{x}_1 = f_1(t,x_1) + g_1(t,x_1) \\ \dot{x}_2 = f_2(t,x_1,x_2) + g_2(t,x_1,x_2)(u + k(t,x_1,x_2,u)), \end{cases} \tag{11}$$

whose nominal system is given by (3) and $k : \mathbb{R}_+ \times \mathbb{R}^p \times \mathbb{R}^q \to \mathbb{R}^q$ is the unknown function.

Our goal is to construct a controller $u(t,x)$ that makes the system (11) globally, uniformly, practically, and exponentially stable. The idea of constructing $u(t,x)$ is to coordinate the control (6), which stabilizes the nominal part (3), and a second control, which corresponds to the term of perturbation.

- ($\mathcal{H}_1'$) The nominal closed-loop system by (6) is globally uniformly practically exponentially stabilized with a Lyapunov function $W(.,.)$ of class C^1 which verifies:

$$c_1\|x\|^p \leq W(t,x) \leq c_2\|x\|^p + c$$
$$\frac{\partial W}{\partial t} + \frac{\partial W}{\partial x}[F(t,x) + G(t,x,\alpha)] \leq -c_3\ \|x\|^p + K_1 e^{-\theta t}$$

where $c_1,\ c_2,\ c_3, c, K_1, \theta, p > 1$ strictly positive constants, with $F(t,x) = \begin{pmatrix} f_1(t,x_1) \\ f_2(t,x_1,x_2) \end{pmatrix}$, $G(t,x,\alpha) = \begin{pmatrix} g_1(t,x_1) \\ g_2(t,x_1,x_2)\alpha(t,x) \end{pmatrix}$.

For all $r_1 > 0$, we propose the following controller:

$$u(t,x) = \alpha(t,x) + u_1(t,x). \tag{12}$$

where $\alpha(t,x)$ is the control stabilizing the nominal system (3) and

$$u_1(t,x) = -\eta(t,x)\frac{\psi(t)}{\|\psi(t)\|\eta(t,x) + r_1}$$

with $\psi^T(t) = \frac{\partial W}{\partial x} g_2(t, x_1, x_2)$ and $\eta(t, x)$ a positive function.

- $(\mathcal{H}_2')$ There exists a positive real function $\rho(t, x)$ such that

$$\begin{aligned} \|k(t, x_1, x_2, u)\| &\le \rho(t, x) + \|u_1\|, \\ \rho(t, x) &< \eta(t, x) \end{aligned} \tag{13}$$

and

$$\int_{t_0}^{t} \rho^2(s, x)\, ds \le M_\rho < +\infty,$$

Under the previous hypotheses, we have the following result:

Theorem 1.5.1. *Consider the perturbed system described by (11) verifying the hypothesis $(\mathcal{H}_1')$ and $(\mathcal{H}_2')$ with the control (12). Then the resulting closed-loop system is globally practically exponentially stable.*

Proof: The idea of the proof is to consider W as a candidate Lyapunov function for the closed-loop system. We then have its derivative along the trajectories of (11) is given by:

$$\begin{aligned} \dot{W}(t, x) &= \frac{\partial W}{\partial t} + \frac{\partial W}{\partial x}[F(t, x) + G(t, x, \alpha)] \\ &+ \frac{\partial W}{\partial x} g_2(t, x_1, x_2)(u_1(t, x) + k(t, x_1, x_2, u(t, x))), \end{aligned}$$

$$\begin{aligned} \dot{W}(t, x) &\le -c_3\|x\|^p + K_1 e^{-\theta t} + \psi^T(t) u_1(t, x) + \psi^T(t) k(t, x_1, x_2, u) \\ &\le -c_3\|x\|^p + K_1 e^{-\theta t} + \psi^T(t) u_1(t, x) + \|\psi(t)\| \|k(t, x_1, x_2, u)\| \\ &\le -c_3\|x\|^p + K_1 e^{-\theta t} + \psi^T(t) u_1(t, x) + \|\psi(t)\|(\rho(t, x) + \|u_1\|) \\ &\le -c_3\|x\|^p + K_1 e^{-\theta t} + \psi^T(t) u_1(t, x) + \|\psi(t)\|(\rho(t, x) + \eta(t, x)). \end{aligned}$$

And then, using the hypothesis $(\mathcal{H}_2')$, it follows that

$$\begin{aligned} \dot{W}(t, x) &\le -c_3\|x\|^p + K_1 e^{-\theta t} + r_1 + \rho(t, x) + \frac{\|\psi(t)\|\rho(t, x) r_1}{\|\psi(t)\|\eta(t, x) + r_1} \\ &\le -c_3\|x\|^p + K_1 e^{-\theta t} + 2r_1 + \rho(t, x). \end{aligned}$$

With $(\mathcal{H}_1')$, we obtain

$$\begin{aligned} \dot{W}(t, x) &\le -\frac{c_3}{c_2}(W(t, x) - c) + K_1 e^{-\theta t} + 2r_1 + \rho(t, x) \\ &\le -\frac{c_3}{c_2} W(t, x) + \frac{c c_3}{c_2} + K_1 e^{-\theta t} + 2r_1 + \rho(t, x). \end{aligned}$$

Let $y(t) = W(t, x) e^{\frac{c_3}{c_2}(t - t_0)}$, then

$$\dot{y}(t) \le (K_1 e^{-\theta t} + 2r_1 + \rho(t, x)) e^{\frac{c_3}{c_2}(t - t_0)}.$$

Integrating between t_0 and t, we obtain $\forall \quad t \ge t_0$,

$$W(t, x) \le W(t_0, x_0) e^{-\frac{c_3}{c_2}(t - t_0)} + \left[\int_{t_0}^{t} \left(\frac{c c_3}{c_2} + K_1 e^{-\theta s} + 2r_1 + \rho(s, x)\right) e^{\frac{c_3}{c_2}(s - t_0)} ds\right] e^{-\frac{c_3}{c_2}(t - t_0)}.$$

Then,

$$\int_{t_0}^{t}\left(\frac{cc_3}{c_2}+2r_1\right)e^{\frac{c_3}{c_2}(s-t_0)}\,ds \;\leq\; \left(c+\frac{2r_1c_2}{c_3}\right)e^{\frac{c_3}{c_2}(t-t_0)},$$

Which implies,

$$\begin{aligned}\int_{t_0}^{t}K_1e^{-\theta s}e^{\frac{c_3}{c_2}(s-t_0)}\,ds &\leq K_1\left(int_{t_0}^{t}e^{-2\theta s}\,ds\right)^{\frac{1}{2}}\left(\int_{t_0}^{t}e^{2\frac{c_3}{c_2}}(s-t_0)\,ds\right)^{\frac{1}{2}}\\ &\leq \frac{K}{2}\sqrt{\frac{c_2}{c_3\theta}}\quad e^{\frac{c_3}{c_2}(t-t_0)}.\end{aligned}$$

Moreover,

$$\begin{aligned}\int_{t_0}^{t}\rho(s,x)e^{\frac{c_3}{c_2}(s-t_0)}\,ds &\leq \left(\int_{t_0}^{t}\rho^2(s,x)\,ds\right)^{\frac{1}{2}}\left(\int_{t_0}^{t}e^{2\frac{c_3}{c_2}(s-t_0)}\,ds\right)^{\frac{1}{2}}\\ &\leq \sqrt{\frac{M_\rho c_2}{2c_3}}\quad e^{\frac{c_3}{c_2}(t-t_0)},\end{aligned}$$

Which implies,

$$W(t,x) \;\leq\; W(t_0,x_0)e^{-\frac{c_3}{c_2}(t-t_0)}+\left[c+\frac{2r_1c_2}{c_3}+\frac{K}{2}\sqrt{\frac{c_2}{c_3\theta}}+\sqrt{\frac{M_\rho c_2}{2c_3}}\right].$$

It follows that,

$$\begin{aligned}\|x\| &\leq \left(\frac{c_2}{c_1}\right)^{\frac{1}{p}}\|x_0\|e^{-\frac{c_3}{pc_2}(t-t_0)}+\left(\frac{c}{c_1}\right)^{\frac{1}{p}}e^{-\frac{c_3}{pc_2}(t-t_0)}\\ &+ \left[\frac{c}{c_1}+\frac{2r_1c_2}{c_3c_1}+\frac{K}{2c_1}\sqrt{\frac{c_2}{c_3\theta}}+\frac{1}{c_1}\sqrt{\frac{M_\rho c_2}{2c_3}}\right]^{\frac{1}{p}}.\end{aligned}$$

Then, for all $t\geq T>0$ (T large enough) and $\varepsilon\in[0,1]$, we have

$$\|x\| \;\leq\; \left(\frac{c_2}{c_1}\right)^{\frac{1}{p}}\|x_0\|e^{-\frac{c_3}{pc_2}(t-t_0)}+(1+\varepsilon)\left[\frac{c}{c_1}+\frac{2r_1c_2}{c_3c_1}+\frac{K}{2c_1}\sqrt{\frac{c_2}{c_3\theta}}+\frac{1}{c_1}\sqrt{\frac{M_\rho c_2}{2c_3}}\right]^{\frac{1}{p}},$$

Thus B_R is globally uniformly exponentially stable with

$$R=(1+\varepsilon)\left[\frac{c}{c_1}+\frac{2r_1c_2}{c_3c_1}+\frac{K}{2c_1}\sqrt{\frac{c_2}{c_3\theta}}+\frac{1}{c_1}\sqrt{\frac{M_\rho c_2}{2c_3}}\right]^{\frac{1}{p}}.$$

It follows that

$$V(t,x) \;\leq\; V(t_0,x_0)e^{-\int_{t_0}^{t}\phi(s)\,ds}+\left[\int_{t_0}^{t}\chi(s)e^{\int_{t_0}^{s}\phi(\tau)\,d\tau}\,ds\right]e^{-\int_{t_0}^{t}\phi(s)\,ds}.$$

So for **case 1**, according to (7) and (8), we will have

$$\begin{aligned} \|x\| \quad &\leq \quad \sqrt{\frac{\omega}{\sigma}}\ \ \|x_0\|\ \ e^{\frac{M_\lambda}{2}}\ e^{-\left(\frac{a_3 a_1 - a_4\gamma}{2a_1}\right)(t-t_0)} \\ &+ \quad \left[\frac{a_1\left(b_4\rho + 4\sqrt{b_1}a_1 r_1\right)}{\sigma b_4(a_1 a_3 - a_4\gamma)} + \frac{c\ a_1}{\sigma}\sqrt{\frac{a_1 M'_\mu}{2(a_1 a_3 - a_4\gamma)}}\right]^{\frac{1}{2}} e^{\frac{M_\lambda}{2}}, \end{aligned}$$

and for **case 2**, by virtue of (7) and (8), we will have

$$\begin{aligned} \|x\| \quad &\leq \quad \sqrt{\frac{\omega}{\sigma}}\ \ \|x_0\|\ \ e^{2a_1(M_\lambda + M_\mu)}\ e^{-\left(\frac{2\sqrt{b_1}b_3 a_1}{b_4}\right)(t-t_0)} \\ &+ \quad \left[\frac{\rho b_4 + 4\sqrt{b_1}a_1 r_1}{4\sigma\sqrt{b_1}b_3 a_1} + \frac{c\ a_1}{\sigma}\left(\frac{b_4 M'_\mu}{8\sqrt{b_1}b_3 a_1}\right)^{\frac{1}{2}}\right]^{\frac{1}{2}} \quad e^{2a_1(M_\lambda + M_\mu)}, \end{aligned}$$

□

and subsequently

$$\begin{aligned} \dot{V}\ (t,x) \quad &\leq \quad -\left(a_3 - \frac{r_1(t)}{a_1} - \frac{b_4\eta\lambda(t)}{2a_1}\right)V_1(t,x_1) - \left(b_3\eta - \frac{b_4\lambda(t)\eta}{2b_1} - \frac{b_4\mu(t)\eta}{\sqrt{b_1}}\right)V_2(t,x_2) \\ &+ \quad \rho + l_1\frac{r_1(t)}{\sqrt{a_1}} + l_2\frac{b_4\eta\mu(t)}{\sqrt{b_1}} + \eta\varepsilon\theta(t,y) + \eta\nu_1\rho(t,y)\|u_0\| \\ &+ \quad \eta\left\|\frac{\partial V_2}{\partial x_2}g_2(t,x_1,x_2)\right\|\|u_0\|\left(-\theta(t,y) - \frac{\nu_1}{\varepsilon}\rho(t,y)\|u_0\|\right). \end{aligned}$$

So, we deduce that

$$\begin{aligned} \dot{V}\ (t,x) \quad &\leq \quad -\left(a_3 - \frac{r_1(t)}{a_1} - \frac{b_4\eta\lambda(t)}{2a_1}\right)V_1(t,x_1) - \left(b_3\eta - \frac{b_4\lambda(t)\eta}{2b_1} - \frac{b_4\mu(t)\eta}{\sqrt{b_1}}\right)V_2(t,x_2) \\ &+ \quad \rho + l_1\frac{r_1(t)}{\sqrt{a_1}} + l_2\frac{b_4\eta\mu(t)}{\sqrt{b_1}}. \end{aligned}$$

Hence, let us choose $\eta = \frac{2b_1}{b_4}$

$$\begin{aligned} \dot{V}\ (t,x) \quad &\leq \quad -\min(\varphi(t),\psi(t))V(t,x) + \tau_1(t) \\ &\leq \quad -\phi(t)V(t,x) + \tau_1(t), \end{aligned}$$

where

$$\begin{aligned} \varphi(t) &= a_3 - \frac{r_1(t)}{a_1} - \frac{b_1}{a_1}\lambda(t),\ \ \psi(t) = \ \frac{2b_1 b_3}{b_4}\eta - \lambda(t) - 2\sqrt{b_1}\mu(t) \\ \tau_1(t) &= \ \rho + l_1\frac{r_1(t)}{\sqrt{a_1}} + l_2\frac{b_4\eta\mu(t)}{\sqrt{b_1}}, \end{aligned}$$

and then, the result follows from theorem 3.1.

6. Global practical stabilization by output loopback

In this section, we consider the following input-output system:

$$\begin{cases} \dot{x}_1 = f_1(t,x_1) + g_1(t,x_1) \\ \dot{x}_2 = f_2(t,x_1,x_2) + g_2(t,x_1,x_2)(u + k(t,x_1,x_2,u)), \\ y = (y_1,y_2) = (x_1,h(t,x_2)) \end{cases} \quad (14)$$

where u, y designate respectively the control (input) and the output of the system, f_1, f_2, h are continuous in t and locally Lipschitzian in x, checking $f_1(t,0) = 0, f_2(t,0,0) = 0$ and $h(t,0) = 0$.

The state $x_2(t)$ is not known; we are then led to the construction of a controller $u(t,y)$ to guarantee the uniform global exponential practical stability of the system (14).

Definition 1.6.1. B_R is globally uniformly exponentially stabilized by the feedback

$$u(t) = u(t,y(t)) \quad (15)$$

if there exists $\gamma > 0$ *and* $k > 0$ *such that for all* $t \geq t_0 \geq 0$ *and* $x_0 \in \mathbb{R}^n$, *the solution* $x(t)$ *of the closed-loop system eqrefsorr-*(15) *checks*:

$$\|x(t)\| \leq k\|x_0\| \, exp\left(-\gamma(t-t_0)\right) + R.$$

In this case, the looped system (14)-(15) *is globally uniformly practically exponentially stable.*

In fact, it is assumed that the following assumption is satisfied.

- $(\mathcal{H}_4)$ There exists a positive real function $\theta_0(t,y)$ such that

$$\|k(t,x_1,x_2,u)\| \leq \theta_0(t,y)\|u\|, \quad (16)$$

for all $t \in \mathbb{R}_+, y \in \mathbb{R}^{p\times q}, u \in \mathbb{R}^q$, and

$$\int_{t_0}^{t} \theta_0^4(t,s)ds \leq M_\theta < +\infty.$$

For all $r_2 > 0$, we propose the following controller

$$u(t,y) = -\frac{g_2^T(t,x_1,x_2)\frac{\partial V_2}{\partial x_2}(t,x_2)\theta_0(t,y)}{\left\|\frac{\partial V_2}{\partial x_2}(t,x_2)g_2,(,tx_1x_2)\right\| + r_2} \quad (17)$$

where V_2 is the function given by the hypothesis $(\mathcal{H}_2)$.

Theorem 1.6.1. *Consider the system (14) and assume that it satisfies the hypotheses* $(\mathcal{H}_1)$, $(\mathcal{H}_2)$ *and* $(\mathcal{H}_4)$. *Then, the closed-loop system (14-17) is globally uniformly practically exponentially stable.*

Proof: Under the hypotheses $(\mathcal{H}_1)$ and $(\mathcal{H}_2)$, we will use, as previously, the function $V(.,.) = V_1(.,.) + \eta V_2(.,.)$ as a candidate Lyapunov function for the closed-loop system. First, we immediately see that

$$\|k(t,x_1,x_2,u)\| \leq \theta_0^2(t,y).$$

and let us differentiate V along the trajectories of (14)

$$\begin{aligned}\dot{V}(t,x) &= \frac{\partial V_1}{\partial t}+\frac{\partial V_1}{\partial x_1}f_1(t,x_1)+\frac{\partial V_1}{\partial x_1}g_1(t,x_1)\\ &+ \eta\frac{\partial V_2}{\partial t}+\frac{\partial V_2}{\partial x_2}f_2(t,x_1,x_2)+\eta\frac{\partial V_2}{\partial x_2}g_2(t,x_1,x_2)u(t,y)\\ &+ \eta\frac{\partial V_2}{\partial x_2}g_2(t,x_1,x_2)k(t,x_1,x_2u)\end{aligned}$$

Moreover, under the hypothesis $(\mathcal{H}_4)$, we have

$$\begin{aligned}\dot{V}(t,x) &\le \frac{\partial V_1}{\partial t}+\frac{\partial V_1}{\partial x_1}f_1(t,x_1)+\frac{\partial V_1}{\partial x_1}g_1(t,x_1)\\ &+ \eta\frac{\partial V_2}{\partial t}+\frac{\partial V_2}{\partial x_2}f_2(t,x_1,x_2)-\eta\frac{\partial V_2}{\partial x_2}g_2(t,x_1,x_2)\frac{g_2^T(t,x_1,x_2)\frac{\partial V_2}{\partial x_2}(t,x_2)\theta_0^2(t,y)}{\left\|\frac{\partial V_2}{\partial x_2}(t,x_2)g_2,(,tx_1x_2)\right\|+r_2}\\ &+ \eta\frac{\partial V_2}{\partial x_2}g_2(t,x_1,x_2)\theta_0^2(t,y),\end{aligned}$$

it follows that,

$$\begin{aligned}\dot{V}(t,x) &\le \frac{\partial V_1}{\partial t}+\frac{\partial V_1}{\partial x_1}f_1(t,x_1)+\frac{\partial V_1}{\partial x_1}g_1(t,x_1)\\ &+ \eta\frac{\partial V_2}{\partial t}+\frac{\partial V_2}{\partial x_2}f_2(t,x_1,x_2)+\eta r_2\theta_0^2(t,y)\end{aligned}$$

We can now proceed as in the proof of Theorem 1.5.1 and take into account the hypotheses $(\mathcal{H}_1)$ and $(\mathcal{H}_2)$. We have

$$\begin{aligned}\dot{V}(t,x) &\le -\left(a_3-\frac{a_4\gamma}{a_1}-\lambda(t)\right)V_1(t,x_1)-\left(\frac{4\sqrt{b_1}b_3a_1}{b_4}-4a_1(\lambda(t)+\mu(t))\right)V_2(t,x_2)\\ &+ \rho+cc_1\mu(t)+\eta r_2\theta_0^2(t,y),\end{aligned}$$

with

$$\begin{aligned}\varphi(t) &= a_3-\frac{a_4\gamma}{a_1}-\lambda(t)\\ \psi(t) &= \frac{4\sqrt{b_1}b_3a_1}{b_4}-4a_1(\lambda(t)+\mu(t))\\ \chi(t) &= \rho+c\ a_1\mu(t)+\eta r_2\theta_0^2(t,y).\end{aligned}$$

It follows that

$$\dot{V}(t,x) \le -\min\{\varphi(t),\psi(t)\}V(t,x)+\chi(t).$$

Then for **case 1**, taking into account (7) and (8), we obtain

$$\begin{aligned} V(t,x) &\le V(t_0,x_0)\ e^{M_\lambda}\ e^{-\left(a_3-\frac{a_4\gamma}{a_1}\right)(t-t_0)} \\ &+ \left[\frac{\rho a_1}{a_1a_3-a_4\gamma} + c\ a_1\sqrt{\frac{a_1M'_\mu}{2(a_1a_3-a_4\gamma)}} + \frac{4a_1r_2}{b_4}\sqrt{\frac{a_1b_1M_{\theta_0}}{2(a_1a_3-a_4\gamma)}}\right] e^{M_\lambda}. \end{aligned}$$

which implies that,

$$\begin{aligned} \|x\| &\le \sqrt{\frac{\omega}{\sigma}}\ \|x_0\|\ e^{\frac{M_\lambda}{2}}\ e^{-\left(\frac{a_1a_3-a_4\gamma}{2a_1}\right)(t-t_0)} \\ &+ \left[\frac{\rho a_1}{\sigma(a_1a_3-a_4\gamma)} + \frac{c\ a_1}{\sigma}\sqrt{\frac{a_1M'_\mu}{2(a_1a_3-a_4\gamma)}} + \frac{4a_1r_2}{\sigma b_4}\sqrt{\frac{a_1b_1M_{\theta_0}}{2(a_1a_3-a_4\gamma)}}\right]^{\frac{1}{2}} e^{\frac{M_\lambda}{2}}. \end{aligned}$$

Then for **case 2**, according to (7) and (8), we obtain

$$\begin{aligned} V(t,x) &\le V(t_0,x_0)e^{4a_1\left(M_\lambda+M_\mu\right)}\ e^{-\left(\frac{4\sqrt{b_1b_3a_1}}{b_4}\right)(t-t_0)} \\ &+ \left[\frac{\rho b_4}{4\sqrt{b_1b_3a_1}} + \left(\frac{b_4}{8\sqrt{b_1b_3a_1}}\right)^{\frac{1}{2}}\left(c\ a_1\sqrt{M'_\mu} + \frac{4a_1r_2}{b_4}\sqrt{b_1M_{\theta_0}}\right)\right] e^{4a_1\left(M_\lambda+M_\mu\right)}, \end{aligned}$$

Thus,

$$\begin{aligned} \|x\| &\le \sqrt{\frac{\omega}{\sigma}}\ \|x_0\|\ e^{2a_1\left(M_\lambda+M_\mu\right)}\ e^{-\left(\frac{2\sqrt{b_1b_3a_1}}{b_4}\right)(t-t_0)} \\ &+ \left[\frac{\rho b_4}{4\sigma\sqrt{b_1b_3a_1}} + \left(\frac{b_4}{8\sqrt{b_1b_3a_1}}\right)^{\frac{1}{2}}\left(\frac{c\ a_1\ b_4\sqrt{M'_\mu} + 4a_1r_2\ \sqrt{b_1M_{\theta_0}}}{\sigma\ b_4}\right)\right]^{\frac{1}{2}} e^{2a_1\left(M_\lambda+M_\mu\right)}, \end{aligned}$$

and subsequently, the looped system (14-17) is globally uniformly exponentially stable. □

7. Conclusion

In this chapter, the problem of output feedback stabilization for nonlinear uncertain systems is investigated. A controller that assures global uniform practical stability of the closed-loop system is proposed; that is, the solutions of the closed-loop system converge toward an arbitrarily small neighborhood of the origin. An illustrative example and simulation are given. It is also shown that the design approach is applicable to more general cases.

Author details

Ines Ellouze and Maryam Ben Salah
University of Sfax, Tunisia

*Address all correspondence to: ines.ellouze@fss.usf.tn

References

[1] Vidyasagar M. Decomposition techniques for large-scale systems with nonadditive interactions stability and stabilisability. IEEE Transactions on Automatic Control. 1980;**25**:773-779

[2] Seibert P, Suarez R. Global stabilization of nonlinear cascade systems. Systems Control Letters. 1990; **14**(4):347-352

[3] Benabdallah A, Hammami MA. On the output feedback stability for non-linear uncertain control systems. International Journal Control. 2001;**74** (6):547-551

[4] Ferfera A, Hammami MA. Growth conditions for global stabilization of cascade nonlinear systems. In: The IFAC Conference Systems Structure and Control. Nantes, France; 1995. pp. 522-526

[5] Hammami MA. Global convergence of a control system by means of an observer. Journal of Optimization Theory and Applications. 2001;**108**(2): 377-388

[6] Saberi A, Kokotovic PV, Sussmann HJ. Global stabilization of partially linear composite systems. SIAM Journal on Control and Optimization. 1990;**28**(6): 1491-1503

[7] Benabdallah A. On the practical output feedback stabilization for nonlinear uncertain systems. Nonlinear Analysis: Modelling and Control. 2009; **14**(2):145-153

[8] Choi HL, Lim JT. Global exponential stabilization of a class of nonlinear systems by output feedback. IEEE Transactions on Automatic Control. 2005;**50**(2):255-257

[9] Jankovic M, Sepulchre R, Kokotovic PV. Constructive Lyapunov stabilization of nonlinear cascade systems. IEEE Transactions on Automatic Control. 1996;**41**(12):1723-1735

Chapter 6

Hybrid Modelling of Water Quality Dynamics: Data Assimilation with Machine Learning for Enhanced Predictions

Parul Tiwari, Channa Rajanayaka and Jing Yang

Abstract

Predicting *Escherichia coli* concentrations in recreational waters is essential for safeguarding public health and ensuring water quality compliance. This study applies time series analysis to forecast *E. coli* levels at six sites in New Zealand using historical data from 2005 to 2020. The goal is to develop a reliable predictive model that helps in proactive water management and early contamination warnings. Initially, an autoregressive integrated moving average (ARIMA) model was applied with parameters selected through a stepwise fitting approach. However, ARIMA demonstrated limitations in accurately capturing *E. coli* variability due to external environmental factors. Then the seasonal autoregressive integrated moving average with exogenous regressors (SARIMAX) model was applied for better predictive performance using water quality parameters and climate variables as input predictors. Results showed that no single water quality parameter consistently predicted *E. coli* across all sites, though total phosphorus emerged as a key predictor in five locations. The four-year forecasts showed patterns aligned with historical trends, suggesting reasonable predictive capability. However, forecast accuracy varied across sites, likely due to site-specific hydrological conditions. This study highlights the importance of site-specific modelling, real-time environmental data integration, and advanced machine learning techniques to improve water quality predictions. A refined forecasting approach can support early warning systems and risk-based decision-making, ultimately reducing health risks associated with microbial contamination in recreational waters.

Keywords: water quality dynamics, climate data, *Escherichia coli*, machine learning, data assimilation

1. Introduction

Water is a natural resource and a priceless gift of nature to all living organisms and is a constant threat of pollution by life itself [1]. It is essential for survival as it carries nutrients to all cells in our body and oxygen to our brain. Clean water is vital to our health, communities, and economy, which is essential to function and flourish and will ultimately lead towards a more sustainable future. Four major categories of water-related diseases are waterborne, water-based, water-related, and water-scarce diseases [2]. Water quality is a complex interplay of physical, chemical, and biological parameters [3]. Several factors are responsible for degradation in water quality [4]. These include the concentration of microscopic pathogens such as bacteria and viruses, quantities of pesticides, insecticides, heavy metals and several other contaminants. Additionally, water quality dynamics are influenced by natural processes such as precipitation, sedimentation, and seasonal temperature variations, as well as anthropogenic activities like wastewater discharge, agricultural runoff, and industrial effluents.

Mathematical models have emerged as indispensable tools for understanding, predicting, and managing water quality [5, 6]. These models provide a quantitative framework for simulating the behaviour of pollutants, assessing the impacts of various stressors and developing management strategies. One widely used model is the advection-diffusion-reaction equation, which describes the transport and transformation of pollutants in water bodies [7]. This model incorporates terms for advection (movement of pollutants with water flow), diffusion (spreading due to concentration gradients), and reaction (chemical and biological transformations). A recent study has applied this model to simulate the dispersion of contaminants in rivers and estuaries, enabling precise predictions of pollution hotspots and their impacts on aquatic ecosystems [8].

The regulatory and advisory committees around the world set up a threshold limit on the microbial quality of water based on the concentrations of *E. coli* [9]. *E. coli*, a type of bacteria commonly found in the intestines of humans and animals, often enters water bodies through sewage discharge, agricultural runoff, or stormwater drainage. When contamination levels are high, water becomes unsafe for public use, posing serious health risks. Exposure to *E. coli*-contaminated water can lead to stomach cramps, diarrhoea, nausea, and even more severe infections [10]. There are various strains of this bacteria. A few of them are enteroaggregative *E. coli* (EAEC), enterohaemorrhagic *E. coli* (EHEC), and enterotoxigenic *E. coli* (ETEC). A specific strain of *E. coli*, called *E. coli* O157:H7, causes severe intestinal infections in humans and can lead to life-threatening conditions like kidney failure [11].

E. coli is ranked third among the 12 antibiotic-resistant priority pathogens identified by the World Health Organisation (WHO). *E. coli* contamination does not just affect human health; it also harms aquatic ecosystems. High levels of *E. coli* often mean there are other pollutants in the water, like nutrients from farms, which can cause algal blooms. These blooms use oxygen in the water, making it hard for fish and other aquatic animals to survive. This chain reaction harms biodiversity, damages ecosystems, and affects the overall health of rivers, lakes, and oceans. Understanding the dynamics of water quality is essential for assessing the state of water bodies and implementing effective management strategies. Accurate prediction of Escherichia coli contamination in surface water is challenging due to considerable uncertainty in the physical, chemical, and biological variables that control *E. coli* occurrence and sources in surface waters [11].

Collecting the samples and then culturing and incubating them until bacteria growth is visible in laboratories takes several hours [12]. Within this time gap, the actual levels of *E. coli* may change notably. For effective water quality management, fast and efficient alternative methods are needed in addition to conventional laboratory methods to estimate the levels of these bacteria. Data-driven predictive models have the capability to collaborate with the hard work of researchers and all contributors in the water industry towards achieving the standards of water quality. Machine learning-based models have the potential to detect and understand unexpected gradual changes in the physical, chemical and biological qualities of water. Based on turbidity, temperature and pH, the authors in [13] applied several machine learning algorithms to estimate water quality and developed a deep neural network with 93% accuracy. The results are tested according to the World Health Organisation (WHO) standards. A combination of single feed-forward and multiple neural networks is used in [14] to get an R^2 and MSE score of 0.9270 and 0.1200, respectively. The authors used 25 features as the input and applied backwards elimination and forward selection combination methods.

The authors in Ref. [15] suggested that the existence of *E. coli* contamination and its prediction in rivers and reservoirs is a major concern because of its adverse effects on human life and food safety. Despite the very low survival rate outside of the host organism of *E. coli*, water resources have often been found to be contaminated [16]. The authors in Ref. [17] found that *E. coli* levels could be predicted using the amounts of algal pigments like chlorophyll and phycocyanin. *E. coli* interacts with algae by giving them organic nutrients and blocking sunlight.

A robust predictive model can be developed with careful consideration of input parameters. It is important to note here that in accounting for all these factors and heterogeneity present in *E. coli* fate and transport, great care is needed to generalise the results. Also, in case of non-point source pollution, concentrations of *E. coli* in waste and manure vary widely, and the dismissal of livestock waste is not well defined. Researchers observed that *E. coli* loads were notably in the high range at high flows compared to low flows [9]. The peaks of turbidity and of *E. coli* loads are compared in [18]. The authors in Ref. [18] observed that *E. coli* peaks are always preceded by turbidity peaks with similar timings, which resulted in a non-linear relationship between *E. coli* and turbidity.

Process-based models using mass conservation principles [19, 20] have been developed. However, hydrodynamic models are effective for water quality dynamics using *E. coli* as a microbial pollutant indicator [21]. A statistical and machine learning model has been developed for predicting *E. coli* loads using water quality, hydrodynamic data and a few other input variables [22]. Several multiple regression models have been developed to predict *E. coli* loads using several input features including turbidity, wave height, and chlorophyll [23–25]. A comparison study between artificial neural network (ANN) and support vector regression (SVR) is done to predict the concentration of *E. coli* at two recreational beaches [26].

In recent years, researchers have applied various data-driven regression models to elaborate the influences of various physicochemical water quality parameters on concentrations of faecal indicator organisms (FIOs) and artificial intelligence methods such as artificial neural network (ANN) and support vector machines [22, 23, 27]. The wide range of applications of machine learning algorithms and computational intelligence as decision support tools has made them indispensable tools [28–30]. Unlike traditional multivariate regression, which assumes variables are independent and models only linear relationships, machine learning uses advanced techniques to find patterns and connections in system data [31, 32]. It can handle complex, non-linear relationships between noisy and interdependent variables. This allows machine learning to make predictions and decisions that are often too complex for traditional methods [33, 34].

Laboratory analysis and the use of artificial intelligence for predictive models enable us to be proactive and to take corrective action for a contaminated site [35]. The applicability of ML methods in water quality management depends on the choice and response of different kernel functions in predictive models [36]. Recreational water activities, such as swimming, kayaking, paddleboarding, and boating, have become increasingly important in today's fast-paced world. These activities provide people with a chance to unwind, reconnect with nature, and engage in physical exercise, which is essential for maintaining good health [37]. Beyond the physical benefits, such activities also play a key role in improving mental well-being, reducing stress, and fostering social connections. Families and friends often bond through shared water experiences, and the sense of calm that comes from being near or on the water can be deeply therapeutic [38]. Additionally, recreational water activities contribute to local economies by attracting tourists and supporting businesses like boat rentals, water parks, and beach resorts.

The aim of this study is to investigate the suitability of different machine learning models to predict the presence of *E. coli* for the New Zealand river dataset as per the recreational and bathing standards. The raw data collected from different sites of rivers in New Zealand provide valuable information in support of utilities to enable them to make better decisions in conjunction with machine learning techniques.

2. Materials and methods

2.1 Site description

This study utilised six different water quality measurement sites across New Zealand – three in the North Island and three in the South Island (**Figure 1**). The North Island sites include Mohaka (HV6), Whanganui (TU1), and Tongariro (TU2), while the South Island sites are Clutha (DN4), Wairau (NN3), and Mataura (DN5).

The Clutha River, the longest in the South Island and the second longest in New Zealand, stretches 338 kilometres (210 miles) from Lake Wānaka in the Southern Alps to the Pacific Ocean, flowing south-southeast through Central and South Otago. The river boasts the largest catchment and outflow in New Zealand, draining a basin of approximately 21,960 square kilometres (8480 square miles) and discharging a mean flow of 614 cubic metres per second.

The Wairau River is in the northeast of the South Island, in the Marlborough region. It originates in the Spenser Mountains and flows northeast to the Cook Strait.

The Mataura River is a significant waterway in the Southland Region of New Zealand. Stretching 240 kilometres (150 miles) from its source in the Eyre Mountains to its mouth at Toetoes Bay on the Pacific Ocean, the Mataura River is renowned for its natural beauty and recreational opportunities. The river faces issues with declining water quality due to intensive farming.

The Tongariro River, located in the North Island of New Zealand, is a vital waterway that flows north from the Central Plateau into Lake Taupō. In its lower reaches, the river's minimum flow ranges from 16 to 21 cubic metres per second. Renowned for its world-class trout fishing, it is often referred to as the "Mecca of trout fishing."

The Mohaka River is a significant waterway in the east-central North Island of New Zealand, spanning 172 km (107 miles) with a catchment area of 2444 square kilometres (910 square miles). Its mean annual flow, measured at Raupunga in the lower catchment, is 78.1 cubic metres per second.

The Whanganui River, one of the longest in the North Island at 290 kilometres (180 miles), flows northwest before turning southwest at Taumarunui, eventually reaching the Tasman Sea. Known for its scenic beauty, it is popular for jet boating, canoeing, and multi-day river journeys.

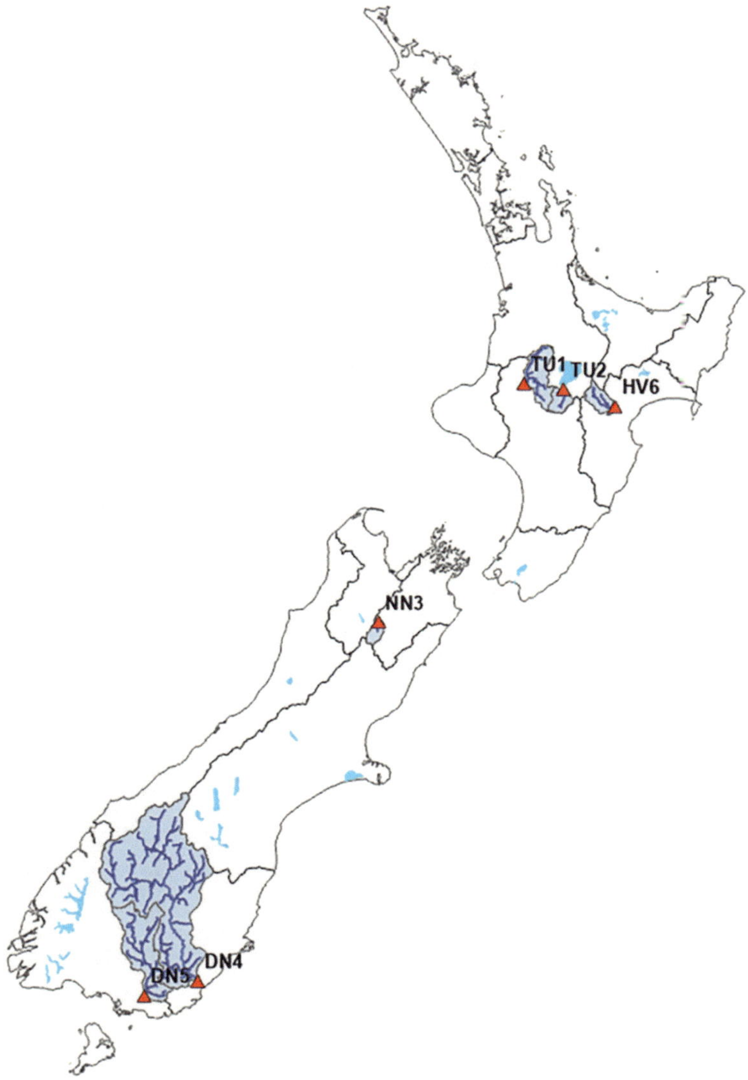

Figure 1.
Location of the sites under study (red triangles: sampling sites; shaded polygons: upstream catchments; and blue lines: upstream river network).

2.2 Data collection

Water quality and streamflow data: Water quality and streamflow data from 1989 to 2020 were obtained from the National Institute for Water and Atmospheric Research's (NIWA) National River Water Quality Network (NRWQN) in New Zealand [39]. This dataset provides extensive information on key physical, chemical, and biological water quality parameters for 35 major rivers, collectively draining approximately half of New Zealand's total land area. Monthly data were collected from 58 active river sites, including 13 water quality variables and two biomonitoring variables, using a combination of *in situ* measurements and grab sampling.

Climate data: Climate data were sourced from NIWA's Virtual Climate Station Network (VCSN) [40], a high-resolution gridded dataset that spans all of New Zealand. The VCSN dataset provides daily weather data including precipitation and temperature at a 5 km x 5 km resolution, offering comprehensive climate coverage across the study area.

Upstream catchment characteristics: Data on upstream catchment characteristics, including climate, geomorphology, geology, and land cover, were obtained from the River Environmental Classification [41] and the Freshwater Ecosystems of New Zealand database [42]. **Table 1** shows the description of sites and the units of water quality and climate variables used in this study.

2.3 Data analysis

Exploratory data analysis (EDA) is performed to understand the distribution of water quality parameters, trends, seasonal patterns, and correlations between *E. coli* levels and other water quality parameters.

Site Description	
Full name	**Abbreviation**
Mataura	DN5
Clutha	DN4
Wairau	NN3
Mohaka	HV6
Whanganui	TU1
Tongariro	TU2
Water Quality and Stream Flow Variables	
Parameter	**Unit**
Clarity (CLAR)	m
Dissolved reactive phosphorus (DRP)	mg/l
Turbidity (TURB)	NTU
Total nitrogen (TN)	mg/l
Total phosphorus (TP)	mg/l
Ammoniacal nitrogen (NH4N)	mg/l
Nitrate-nitrite-nitrogen (NNN)	mg/l
Escherichia coli (ECOLI)	cfu/100 ml
Climate Variables	
Parameter	**units**
Daily flow	m^3/s
Rainfall	mm
Temperature	degC

Table 1.
Description of the sites and input parameters.

The mean concentration of *E. coli* is higher at sites DN5 and DN4 than the other four sites. This indicates that these locations have consistently elevated *E. coli* levels, which may suggest a persistent contamination source, such as sewage discharge, agricultural runoff, or reduced water flow leading to stagnation. Descriptive statistics of *E. coli* values (in cfu/100 ml) for all six sites are given in **Table 2**.

Sites DN5 and TU1 also show a higher standard deviation in *E. coli* concentrations. A higher standard deviation means that the *E. coli* levels at these sites fluctuate more significantly over time. This variability could be due to irregular contamination events, seasonal changes, varying pollution sources, or differences in water dynamics affecting bacterial distribution. Among all the sites, NN3 has the lowest mean concentration of *E. coli*, and its maximum recorded value is also the lowest compared to the other sites. This suggests that NN3 is the least contaminated site, likely benefiting from cleaner water sources, better dilution, or effective natural self-purification processes.

The density plots for these sites provide a visual representation of the distribution of *E. coli* concentrations and help to further interpret the statistical findings. Sites DN5 (Mataura), DN4 (Mohaka), and TU1 (Whanganui) frequently experience higher levels of contamination. In contrast, the density plot for NN3 (Wairau) is much narrower and concentrated at lower *E. coli* values, reflecting its lowest mean concentration and the fact that even its maximum recorded value is lower than that of the other sites.

Additionally, the density plots for DN5 and TU1 likely exhibit a wider spread, indicating a greater variability in *E. coli* concentrations. This broader distribution corresponds with the higher standard deviation observed at these sites, suggesting that their *E. coli* levels fluctuate more significantly over time. The density plots showing the distribution of *E. coli* values are plotted in **Figure 2**.

We calculated skewness and kurtosis to analyse the variability in data across all sites. **Table 3** shows these statistical measures across all sites. The skewness and kurtosis values for Wairau are the highest among all the sites. This high value suggests that its distribution is heavily skewed, likely due to the presence of very low values with occasional slightly higher readings. This indicates that *E. coli* concentrations at NN3 are generally low, with a few isolated instances of relatively higher values.

Statistic	Site name					
	DN5	DN4	NN3	HV6	TU2	TU1
Count	185	185	185	160	185	185
Mean	979.27	197.74	23.73	90.31	55.47	336.68
Std	2149.34	471.26	138.50	324.28	182.56	552.29
Min	2	2	0	1	0	0
25%	172.2	18	1	9.7	10.9	59.8
50%	359	37.9	3.2	17.7	17	114.5
75%	878	90.9	9	37.75	32.3	365.4
Max	17,328	2613	1597	3076	1935	3255

Table 2.
Descriptive measures for E. coli *values across six sites under study.*

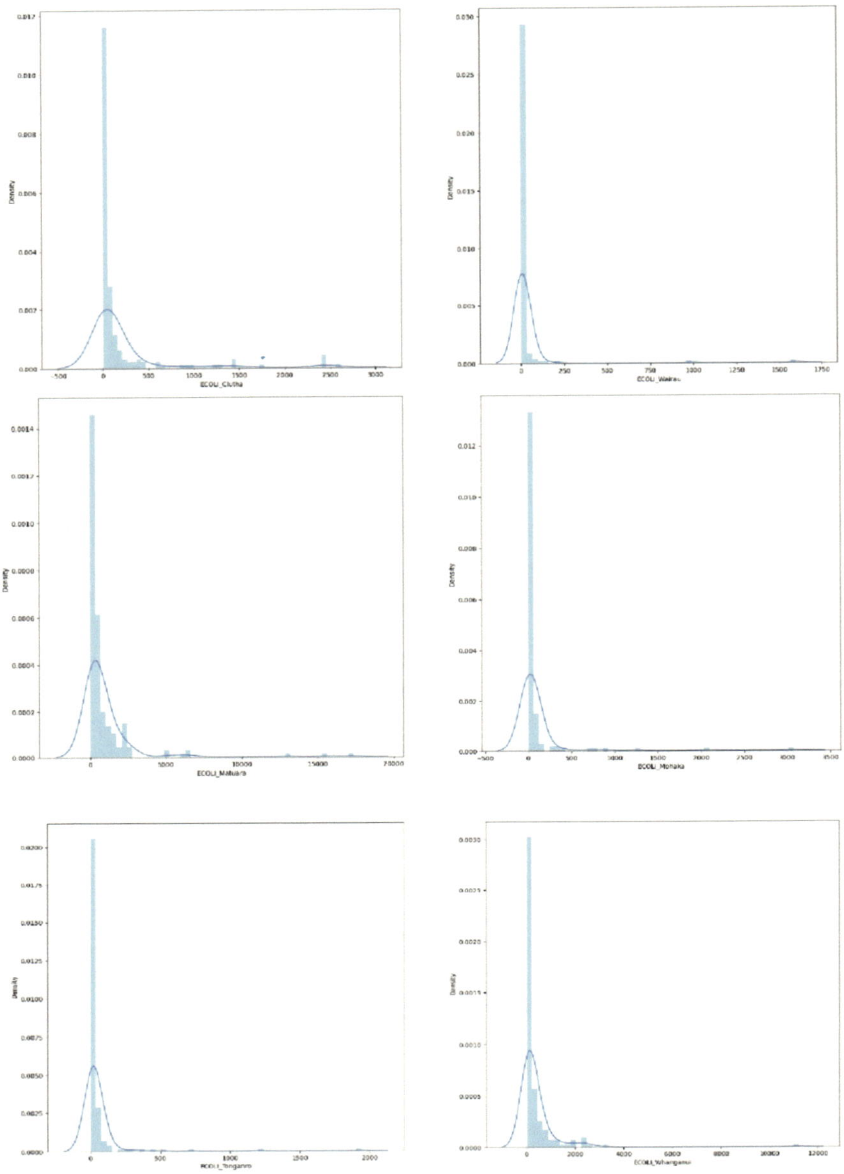

Figure 2.
Density plots for E. coli *values across different sites.*

Statistical measures	Sites					
	Matauara	**Clutha**	**Wairau**	**Mohaka**	**Tongariro**	**Whanganui**
Skewness	5.46	3.68	9.92	6.91	7.83	8.12
Kurtosis	34.44	13.70	103.61	54.38	70.62	85.39

Table 3.
Statistical analysis of data variability across sites.

Kurtosis, on the other hand, measures the "tailedness" of the distribution. A higher kurtosis value for Wairau means that its *E. coli* concentration data is more peaked with heavier tails. This suggests that most of the readings are clustered around very low values, with fewer occurrences of extreme values compared to other sites. The high kurtosis also indicates that *E. coli* levels at this site are more stable, with rare fluctuations in contamination levels.

We also plotted facets of input water quality parameters to visualise and compare trends across different monitoring sites. The facets for all input water quality parameters including *E. coli* for the Clutha River are shown in **Figure 3**.

All sites have variations in water quality parameters. *E. coli* values in some of the sites (Wairau and Tongariro) are very low compared to other sites (Mataura and Clutha). **Figure 4** shows the distribution of *E. coli* values (cfu/100 ml) for all six sites. It is clear from the graph that each site bears very high peaks. It could be due to unexpected or sudden changes in some parameters. To study this, we performed a time series analysis for all the sites.

We created box plots for all sites to visually analyse the distribution, spread, and potential outliers in the data. Box plots help us identify the central tendency, variability, and skewness of the data while highlighting any extreme values. This visualisation allows for a quick comparison across different sites, ensuring a deeper understanding of data consistency and detecting any anomalies or irregular patterns. Using box plots, we observed that site DN5 (Mataura) has high variability and includes a

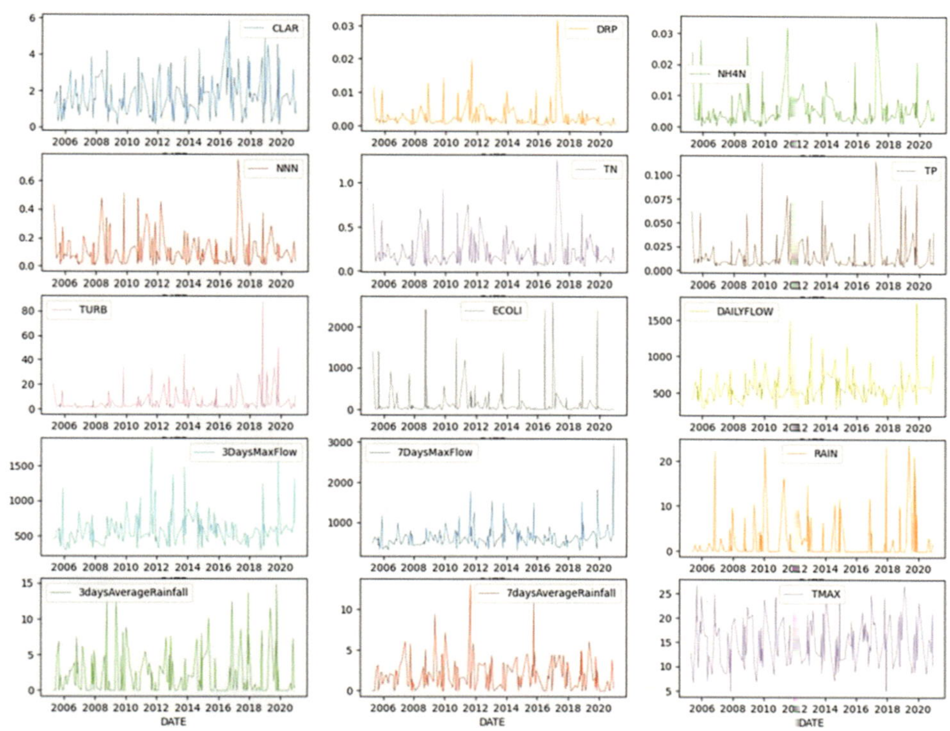

Figure 3.
Facets for water quality parameters for the Clutha river. Units of the variables are given in **Table 1**.

maximum number of outliers. **Figure 5** illustrates that values of turbidity, *E. coli* and total phosphorus have more variability in river Mataura.

Karl Pearson's coefficient of correlation is calculated to observe the relationship between water quality parameters. Correlation heatmaps for the sites are shown in **Figure 6**.

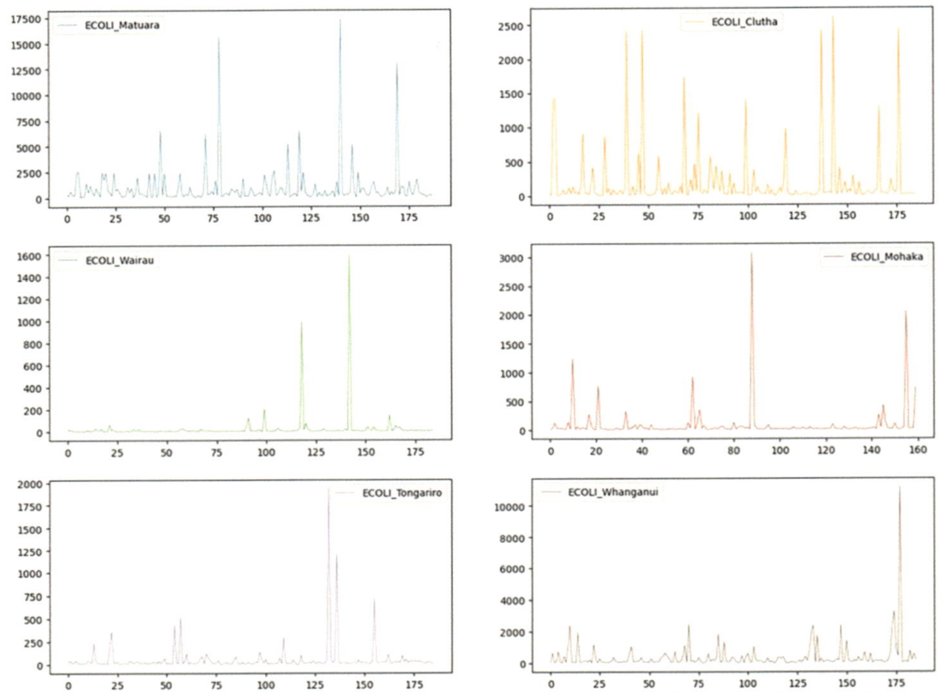

Figure 4.
Monthly E. coli *concentrations (cfu/100 ml) at six selected sites.*

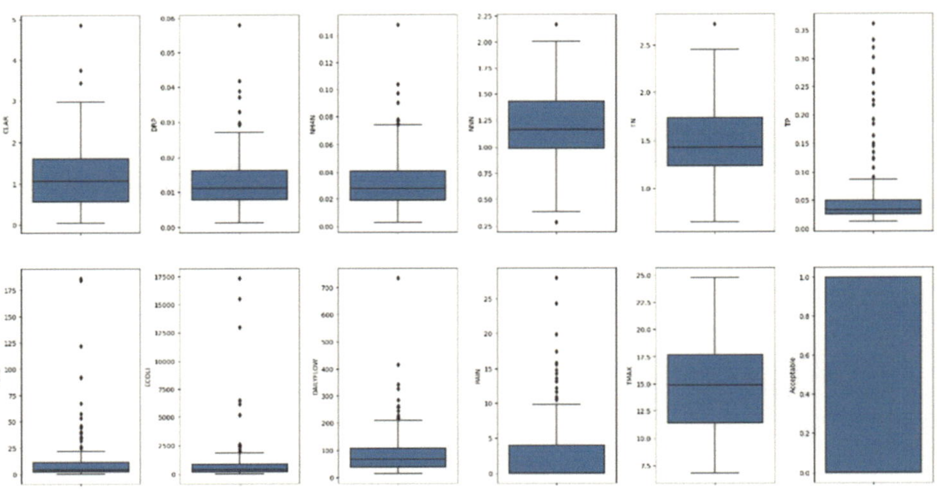

Figure 5.
Box plot for water quality and climate data distribution in the Mataura river.

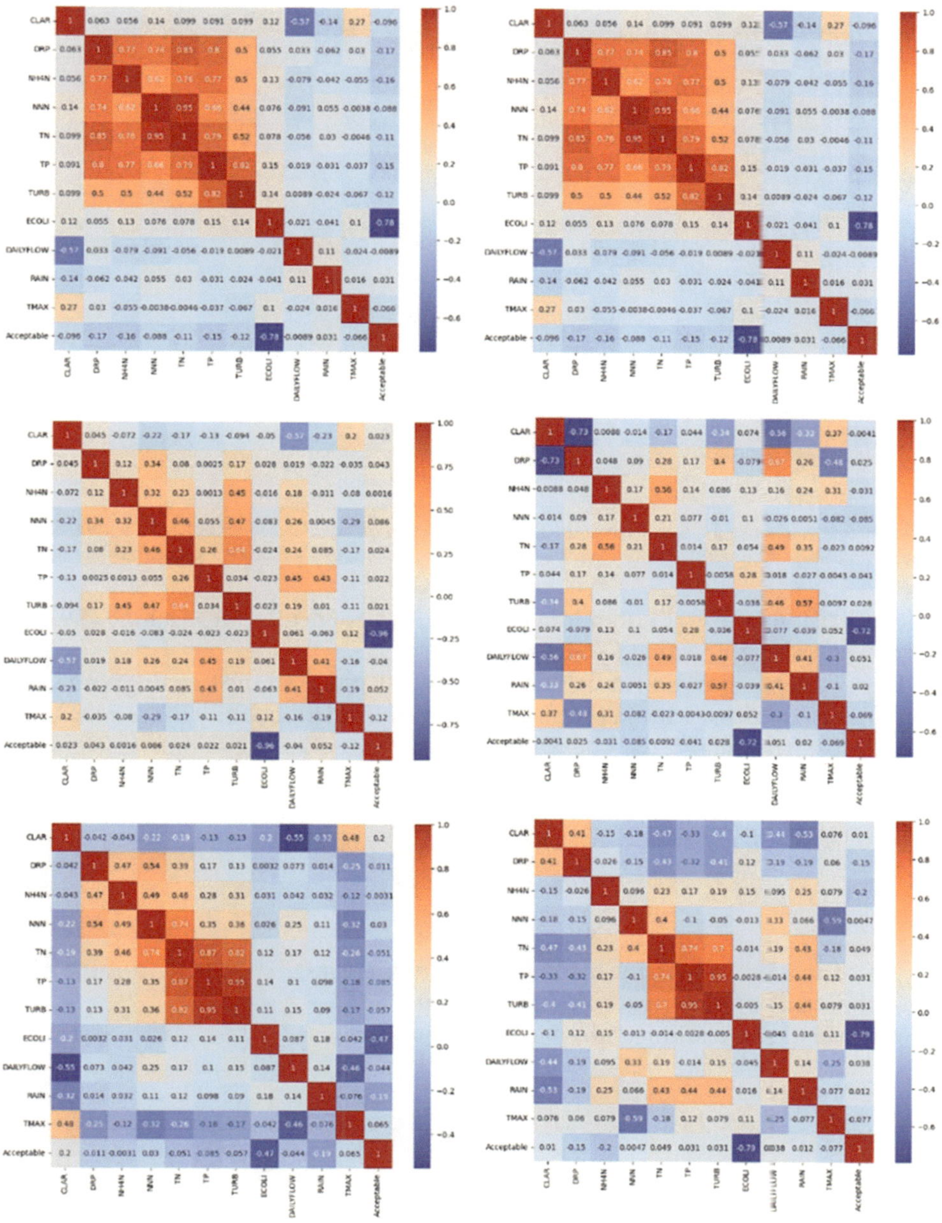

Figure 6.
Heat map for all sites using Pearson's correlation coefficient of the observed water quality parameters.

Land Air Water Aotearoa (LAWA) displays the results of water contamination regularly at recreational sites. **Figure 7** shows the acceptable threshold value of *E. coli* for recreational activities suggested by LAWA.

2.4 Classification algorithms

To check the suitability of water quality for recreational and bathing activities for these sites, we applied five classification algorithms including Logistic Regression

Mode	Trigger level		Management response
	Beach: Enterococci / 100mL	**River/Lake: *E. coli*/100 mL**	
Surveillance	Equal to or less than 140 Enterococci / 100 mL	Equal to or less than 260 *E. coli* / 100 mL	Routine monitoring.
Alert	More than 140 Enterococci / 100 mL	More than 260 *E. coli* / 100 mL	Increase monitoring and investigate source.
Action	More than 280 Enterococci / 100 mL	More than 540 *E. coli* / 100 mL	Public warnings if required, increased monitoring and investigation of contaminant source.

Figure 7.
Standard threshold value for E. coli *concentration suitable for recreational activities in New Zealand. Source: https://www.lawa.org.nz/learn/factsheets/can-i-swim-here/coastal-and-freshwater-recreational-monitoring.*

(LR), Decision Tree (DT), Random Forest (RF), Support Vector Machine (SVM), and Extreme Gradient Boosting (XGBoost).

Logistic regression is a fundamental statistical and machine learning technique used for binary classification problems. It predicts the probability that an instance belongs to a particular class. The core of logistic regression is the logistic (sigmoid) function:

$$\sigma(z) = \frac{1}{1+e^{-Z}} \tag{1}$$

where z is the input (linear combination of features), e is the base of the natural logarithm, and $\sigma(z)$ outputs a value between 0 and 1.

The logistic regression model can be represented as:

$$P(y=1|x) = \sigma\left(w^0 + w^1x^1 + w^2x^2 + \ldots + w_nx_n\right) \tag{2}$$

where $P(y=1|x)$ is the probability that $y=1$ given features x. Here w_0 is the bias term (intercept), $w_1, w_2, \ldots, w_n$ are the weights and $x_1, x_2, \ldots, x_n$ are the feature values. The Gradient Descent method is used for optimisation. The weights are updated using gradient descent as

$$w := w - \frac{\alpha \partial J(w)}{\partial w} \tag{3}$$

where α is the learning rate and $\partial C(w)/\partial w$ is the gradient of the cost function $C(w)$. The learning rate (α) is crucial for successful training. Larger values of (α) may overshoot the minimum values of the function, whereas much smaller values will result in slow convergence. We chose adaptive learning rates in which we started with a larger learning rate of 0.1 and decreased the rate to 0.01 as training progressed.

In the Decision Tree classifier, we used the Classification and Regression Tree (CART) Gini index to achieve maximum information.

Gini index $G(T)$ measures the impurity of a set of classes and is calculated as

$$G(T)=1-\sum_{1}^{c} p_i^2 \quad (4)$$

where T is the current state and p_i is the probability of class i.

Random forest algorithm aggregates several decision trees to form predictions. This involves constructing a multitude of decision trees, where each tree is trained on a random collection of the training dataset, and a random selection of features is used to split the data at every level. When predicting, all the trees in the forest provide their output, and the average (in case of regression) or majority vote (in case of classification) is taken.

Mathematically, if we have N decision trees, the final prediction for classification is given by:

$$\hat{y}=\text{mode}\{T_1(x),T_2(x),\ldots,T_N(x)\} \quad (5)$$

where $T_i(x)$ represents the prediction from the i th tree. For regression, the prediction is computed as

$$\hat{y}=\frac{1}{N}\sum_{i=1}^{N} T_i(x) \quad (6)$$

where the predictions from all trees are averaged.

Based on the classification report, the total number of acceptable (good for recreational activities) and not acceptable (not good for recreational activities) data points for the six sites are shown in **Figure 8**.

Table 4 presents model selection results using five classification algorithms. The RF algorithm and XGBoost produced the best performance across all metrics. These models are superior in performance metrics including accuracy, F1 score, Precision and Recall. Another ensemble method, DT, provided comparable performance with an accuracy of 98.41%. The remaining two algorithms, LR and SVM, show good predictive performances but lower performance than the other algorithms.

2.5 Model development

2.5.1 Time series analysis

Time series analysis was conducted to examine the temporal patterns, trends, and residual variations in the *E. coli* values across all sites. This analysis allowed us to decompose the data into three key components: trend, seasonality, and residuals. Trend shows the long-term trend in the data, indicating whether values increase, decrease, or remain stable over time. Seasonality focuses on repeating patterns

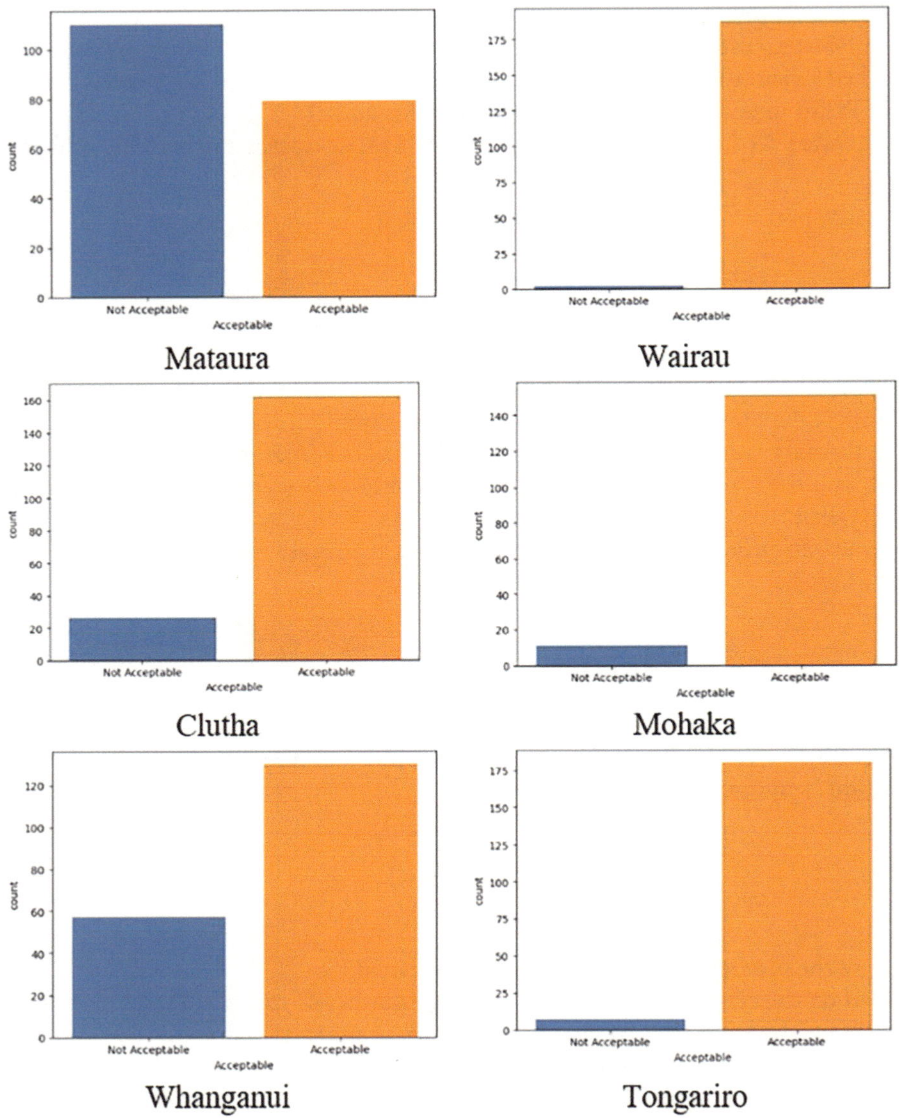

Figure 8.
Bar plot for classification results based on E. coli *concentration present across all sites.*

observed at regular intervals, suggesting periodic fluctuations in the data, and residuals capture the irregular variations that remain after accounting for trend and seasonal effects, helping us assess noise and unexpected changes.

Trends were plotted from year 2005 to year 2020. **Figure 9** illustrates the temporal variations in *E. coli* values across all six sites. From the analysis, we found that seasonal patterns were evident, indicating that the data includes recurring fluctuations at specific time intervals. This suggests that external factors, such as temperature, rainfall and river flow, played a significant role in shaping the data trends. The residuals, while present, showed no dominant pattern, confirming that the primary variations were well captured by the trend and seasonal components. Identifying seasonality helped in predicting future patterns, which is particularly useful for planning and forecasting.

Classification algorithm	Prediction	Precision	Recall	F1-score	Accuracy
Logistic Regression (LR)	0	0.80	0.89	0.84	0.8095
	1	0.83	0.70	0.76	
Decision Tree (DT)	0	1.00	0.97	0.99	0.9841
	1	0.96	1.00	0.98	
Random Forest (RF)	0	1.00	1.00	1.00	1.00
	1	1.00	1.00	1.00	
SVM	0	0.69	0.81	0.74	0.6825
	1	0.67	0.52	0.58	
XGBoost	0	1.00	1.00	1.00	1.00
	1	1.00	1.00	1.00	

Table 4.
Performances of different classification algorithms for E. coli *concentration.*

E. coli concentrations showed a clear seasonal variation, with higher values observed during warmer months, likely due to increased bacterial growth, runoff, and recreational water use. The three sites, Mataura, Clutha, and Whanganui, have high residuals compared to other sites, which shows that these sites have experienced unexpected changes. These changes are observed due to the past few days (3 days) max flow in river during those instances.

2.5.2 Stationarity analysis

We used the Augmented Dickey-Fuller (ADF) test to determine stationarity. A stationary time series has constant mean, variance, and autocovariance over time, making it easier to model and forecast. The test showed that all the time series under study are stationary. We also examined Autocorrelation function (ACF) and Partial Autocorrelation Function (PACF) as shown in **Figure 10**. The ACF plot helped us identify whether past values influenced future values and to what extent. High autocorrelation at certain lags suggested seasonality in *E. coli* values and helped us in choosing the model parameters. In addition, the PACF plot helped us understand the direct relationship between observations at different lags, excluding the influence of intermediate points. This was particularly useful for selecting the appropriate order for autoregressive (AR) components in our model.

The site Mohaka has no lags, which means that past *E. coli* values at this location do not significantly influence future values. This is due to high variability in rainfall and temperature data.

2.5.3 ARIMA model

For model development, we chose the ARIMA model initially. We used the stepwise fit function to determine the optimal parameters for the ARIMA model. These parameters varied for different sites due to differences in local environmental conditions, pollution sources, hydrological patterns, and data trends unique to each location. Despite applying the ARIMA model, the forecasting results were not highly accurate. This could be attributed to several factors, including the high variability of *E. coli* levels, external

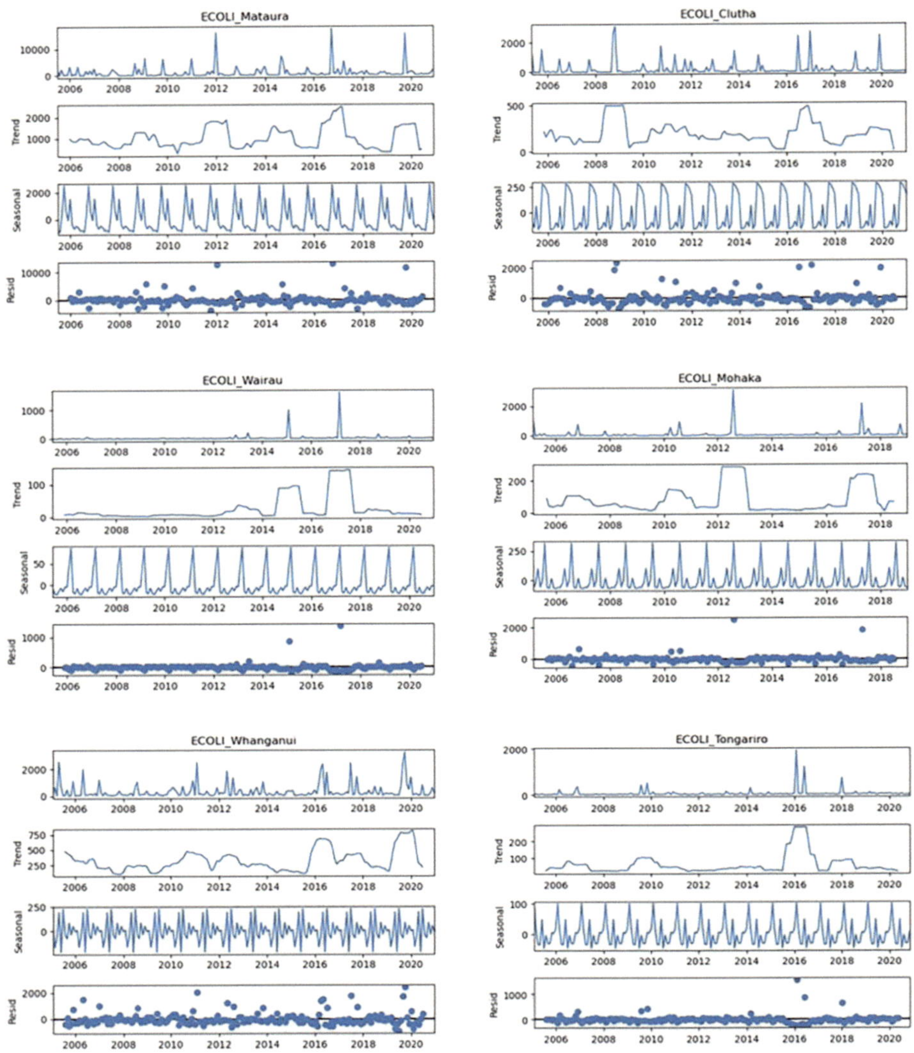

Figure 9.
Time series decomposition of E. coli *concentration across all sites.*

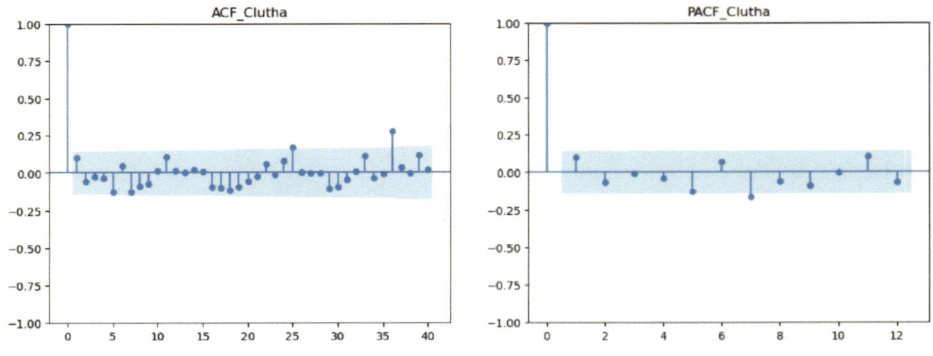

Figure 10.
Autocorrelation function (ACF) and partial autocorrelation function (PACF) for the Clutha river.

environmental influences such as rainfall and temperature variations, and possible data limitations. Additionally, *E. coli* concentrations may be influenced by non-linear and complex interactions that ARIMA, a linear model, may not fully capture.

2.5.4 SARIMAX model

Since the ARIMA model did not provide accurate results, we implemented the Seasonal Auto Regressive Integrated Moving Average with Exogenous variables (SARIMAX) model. SARIMAX extends ARIMA by incorporating seasonality and external regressors, making it more effective in capturing environmental influences on *E. coli* levels. We used other water quality parameters such as rainfall, 3-day average rainfall, total phosphorus, total nitrogen, turbidity, as an external regressor and found that no one water quality parameter accurately predicts *E. coli* levels. It is different for each site, but total phosphorus is common for the five sites except Whanganui. These external variables can help improve the accuracy of forecasts by accounting for additional information that might impact the time series. SARIMAX models combine the concepts of SARIMA models with the ability to include exogenous variables. The main components of SARIMAX models are as follows:

Seasonal Autoregressive (SAR) Component (P): Captures the relationship between the current value and past values at the same seasonal lag.

Seasonal Integrated (SI) Component: Involves differencing the series at the seasonal interval to achieve seasonal stationarity.

Seasonal Moving Average (SMA) Component (Q): Models the relationship between the current value and past error terms at the same seasonal *lag*.

Exogenous (X) Variables: These are external factors that might influence the time series. Including exogenous variables helps the model capture additional patterns and relationships beyond the inherent time series components.

SARIMAX model is represented as $SARIMAX\left(p, d, q\right)\times\left(P, D, Q, s\right)$, where p is the order of the non-seasonal AutoRegressive (AR) component, d is the degree of non-seasonal differencing, q is the order of the non-seasonal moving average (MA) component. In addition, D is the degree of seasonal differencing and s is the number of time steps in each seasonal period. Fitting a SARIMAX model with exogenous variables involves not only identifying the appropriate orders for the seasonal and non-seasonal components but also selecting and incorporating relevant exogenous variables.

3. Results and discussion

Using time series analysis, we predicted *E. coli* concentrations for the next 4 years based on historical data from 2005 to 2020. The predictions provided valuable insights into future trends. In this analysis, we found that total phosphorus (TP) serves as an exogenous variable for all sites except the Mohaka, for which 3-day average rainfall is the exogenous variable. The predicted *E. coli* values followed similar seasonal and long-term patterns observed in the historical dataset. This indicates that our model successfully captured the periodic nature of *E. coli* fluctuations, including seasonal peaks and dips. The model treated the extreme values as outliers. Predicted *E. coli* values along with historic data are plotted in **Figure 11**.

Sites with seasonal variations in past data showed similar trends in future projections, reinforcing the model's effectiveness in detecting cyclic behaviour. The

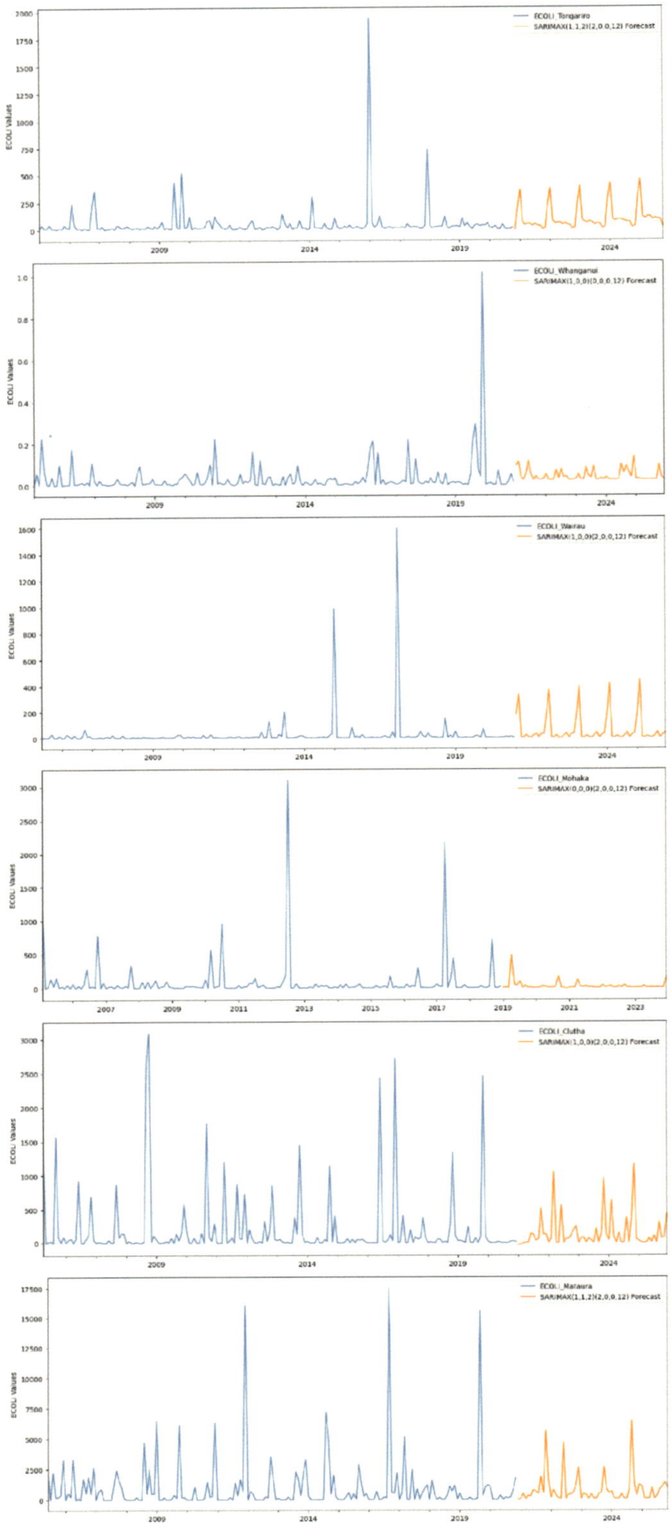

Figure 11.
Forecasting for E. coli *concentration across all sites from year 2021 to year 2024.*

forecasted values were evaluated using statistical metrics, Mean Absolute Error (MAE) and Root Mean Square Error (RMSE). These metrics indicated an accuracy of 86% for the model, confirming that this model effectively captured the underlying patterns of *E. coli* variations. Cross-validation with historical data showed that the model was able to reproduce past trends with minimal error. Furthermore, the projected *E. coli* concentrations did not show unrealistic spikes, suggesting that the model did not overfit to noise in the data. The predicted values remained within the expected range observed in past records, making them more reliable for practical use in water quality monitoring and risk assessment.

However, an important point to note here is that while overall trends were consistent, the rate of increase or decrease in *E. coli* concentrations varied across different sites. This highlights the influence of site-specific environmental factors, which are rainfall, temperature, river flow and chemical variables such as TN, TP, and NH4N. The Clutha River and the Whanganui River showed greater variability, while others exhibited more stable forecasts, suggesting the need for localised calibration and continuous monitoring.

4. Limitations

Traditional ARIMA and SARIMAX models primarily rely on historical patterns and assume that future trends will follow similar behaviour, which may not always hold true in highly variable environmental systems such as heavy rainfall, land usage and pollution events. *E. coli* concentrations in natural water bodies are affected by multiple external factors, including temperature, precipitation, agricultural runoff, and wastewater discharge. Although we incorporated temperature, predictions were not consistent across all sites.

5. Conclusion

This study provided valuable insights into predicting *E. coli* concentrations across six sites in New Zealand. By analysing historical data from 2005 to 2020, we applied ARIMA and SARIMAX models to forecast *E. coli* levels for the next 4 years. These findings demonstrated that while the models captured overall trends and seasonal variations, their accuracy was site-dependent, and external environmental factors played a significant role in influencing contamination patterns.

The presence of elevated *E. coli* levels in recreational waters poses health risks, as it indicates possible faecal contamination, which can lead to waterborne illnesses. Reliable forecasting models can help authorities implement timely interventions, such as issuing public health advisories, monitoring pollution sources, and mitigating contamination risks. This study highlighted some key challenges.

No single water quality parameter, except total phosphorus and 3-day average rainfall, could consistently predict *E. coli* levels across all sites, indicating the complexity of microbial contamination. While historical trends were useful in forecasting, external factors, such as rainfall and temperature, introduced significant variability. Site-specific differences in *E. coli* behaviour suggest that a one-size-fits-all approach is not ideal for water quality forecasting. More localised models or regional calibration techniques may be necessary to improve forecast reliability for specific locations. Regional councils may utilise real-time sensor technology and remote sensing data to capture rapid changes in water quality, and the frequency of water sampling can be increased to improve model resolution.

Future research should aim to enhance predictive models by integrating real-time environmental data, using advanced modelling techniques, and tailoring approaches to site-specific conditions. These improvements will strengthen forecasting accuracy, making it a valuable tool for safeguarding recreational water quality and minimising health risks from waterborne pathogens.

Conflict of interest

The authors declare no conflict of interest.

Author details

Parul Tiwari[1*], Channa Rajanayaka[2] and Jing Yang[2]

1 Department of Mathematical Sciences, Auckland University of Technology, New Zealand

2 National Institute of Water and Atmospheric Research, Christchurch, New Zealand

*Address all correspondence to: parul.tiwari@aut.ac.nz

References

[1] He C, Liu Z, Wu J, Pan X, Fang Z, Li J, et al. Future global urban water scarcity and potential solutions. Nature Communications. 2021;**12**:1. Available from: https://www.nature.com/articles/s41467-021-25026-3

[2] Australian Drinking Water Guidelines. Version 3.8. Australia: National Health and Medical Research Council (NHMRC); Sept 2022. Available from: https://www.nhmrc.gov.au/about-us/publications/australian-drinking-water-guidelines

[3] Kumar D, Kumar R, Sharma M, Awasthi A, Kumar M. Global water quality indices: Development, implications, and limitations. Total Environment Advances. 2024;**9**:200095

[4] World Health Organization. Guidelines for safe recreational water. Volume 1. Coastal and fresh waters. Geneva. 2003;**1**:219. Available from: http://www.who.int/water_sanitation_health/bathing/srwe2full.pdf

[5] Ziemińska-Stolarska A, Skrzypski J. Review of mathematical models of water quality. Ecological Chemistry and Engineering S. 2012;**19**(2):197-211

[6] Tiwari PK, Singh RK, Khajanchi S, Kang Y, Misra AK, Tiwari PK, et al. A mathematical model to restore water quality in urban lakes using Phoslock. Discrete and Continuous Dynamical Systems – B. 2021;**26**(6):3143-3175. Available from: https://www.aimsciences.org/en/article/doi/10.3934/dcdsb.2020223

[7] Yadav RR, Roy J. Analytical solutions of one-dimensional scale dependent advection-dispersion equations for finite domain solute transport. Groundwater for Sustainable Development. 2022;**16**:100712

[8] Sanskrityayn A, Suk H, Chen JS, Park E. Generalized analytical solutions of the advection-dispersion equation with variable flow and transport coefficients. Sustainability. 2021;**13**(14):7796. Available from: https://www.mdpi.com/2071-1050/13/14/7796/htm

[9] Abimbola OP, Mittelstet AR, Messer TL, Berry ED, Bartelt-Hunt SL, Hansen SP. Predicting *Escherichia coli* loads in cascading dams with machine learning: An integration of hydrometeorology, animal density and grazing pattern. Science of The Total Environment. 2020;**722**:137894

[10] Kapinusova G, Lopez Marin MA, Uhlik O. Reaching unreachables: Obstacles and successes of microbial cultivation and their reasons. Frontiers in Microbiology. 2023;**14**:1089630. Available from: https://pmc.ncbi.nlm.nih.gov/articles/PMC10027941/

[11] Ahmed U, Mumtaz R, Anwar H, Shah AA, Irfan R, García-Nieto J. Efficient water quality prediction using supervised machine learning. Water. 24 Oct 2019;**11**(11):2210, 14

[12] Wen X, Chen F, Lin Y, Zhu H, Yuan F, Kuang D, et al. Microbial indicators and their use for monitoring drinking water quality—A review. Sustainability. 2020;**12**(6):2249. Available from: https://www.mdpi.com/2071-1050/12/6/2249/htm

[13] Ebomah KE, Adefisoye MA, Okoh AI. Pathogenic *Escherichia coli* strains recovered from selected aquatic resources in the Eastern Cape,

South Africa, and its significance to public health. International Journal of Environmental Research and Public Health. 2018;**15**(7). Available from: https://pubmed.ncbi.nlm.nih.gov/30018212/

[14] Delair Z, Schoeman M, Reyneke B, Singh A, Barnard TG. Assessing the impact of *Escherichia coli* on recreational water safety using quantitative microbial risk assessment. Journal of Water and Health. 2024;**22**(10):1781-1793. Available from: http://iwaponline.com/jwh/article-pdf/22/10/1781/1499046/jwh2024081.pdf

[15] Efting AA, Snow DD, Fritz SC. Cyanobacteria and microcystin in the Nebraska (USA) Sand Hills lakes before and after modern agriculture. Journal of Paleolimnology. 2011;**46**(1):17-27

[16] Pandey PK, Kass PH, Soupir ML, Biswas S, Singh VP. Contamination of water resources by pathogenic bacteria. AMB Express. 2014;**4**(1):1-16. Available from: https://amb-express.springeropen.com/articles/10.1186/s13568-014-0051-x

[17] Stocker MD, Pachepsky YA, Hill RL. Prediction of *E. Coli* concentrations in agricultural pond waters: Application and comparison of machine learning algorithms. Front. Artificial Intelligence. 2022;**4**:768650. Available from: www.frontiersin.org

[18] Dwivedi D, Mohanty BP, Lesikar BJ. Estimating *Escherichia coli* loads in streams based on various physical, chemical, and biological factors. Water Resources Research. 2013;**49**(5):2896. Available from: https://pmc.ncbi.nlm.nih.gov/articles/PMC3914718/

[19] Baffaut C, Benson VW. A bacteria TMDL for Shoal Creek using SWAT modeling and DNA source tracking. In: Total Maximum Daily Load (TMDL) Environmental Regulations II. American Society of Agricultural and Biological Engineers; 2003. p. 1

[20] Coffey R, Benham B, Wolfe ML, Cummins E. Predicting the effects of environmental change on microbial transport. In: ASABE 1st Climate Change Symposium: Adaptation and Mitigation. 2015. pp. 219-221

[21] Eregno FE, Tryland I, Tjomsland T, Kempa M, Heistad A. Hydrodynamic modelling of recreational water quality using *Escherichia coli* as an indicator of microbial contamination. Journal of Hydrology. 2018;**561**:179-186

[22] Kuroki S, Ogata R, Sakamoto M. Predicting the presence of *E. coli* in tap water using machine learning in Nepal. Water and Environment Journal. Aug 2023;**37**(3):402-411

[23] Cyterski M, Shanks OC, Wanjugi P, McMinn B, Korajkic A, Oshima K, et al. Bacterial and viral fecal indicator predictive modeling at three Great Lakes recreational beach sites. Water Research. 2022;**223**:118970. Available from: https://pmc.ncbi.nlm.nih.gov/articles/PMC9724166/

[24] Francy DS, Stelzer EA, Duris JW, Brady AMG, Harrison JH, Johnson HE, et al. Predictive models for Escherichia coli concentrations at inland lake beaches and relationship of model variables to pathogen detection. Applied and Environmental Microbiology. 2013;**79**(5):1676-1688

[25] Herrig IM, Böer SI, Brennholt N, Manz W. Development of multiple linear regression models as predictive tools for fecal indicator concentrations in a stretch of the lower Lahn River, Germany. Water Research. 2015;**85**:148-157. Available from: https://pubmed.ncbi.nlm.nih.gov/26318647/

[26] Park Y, Cho KH, Park J, Cha SM, Kim JH. Development of early-warning protocol for predicting chlorophyll-a concentration using machine learning models in freshwater and estuarine reservoirs, Korea. Science of The Total Environment. 2015;**502**:31-41

[27] van der Meulen ES, Tertienko A, Blauw AN, Sutton NB, van de Ven FHM, Rijnaarts HHM, et al. A review of prediction models for *E. Coli* in urban surface waters. Urban Water Journal. 2024;**21**(5):539-548. Available from: https://www.tandfonline.com/doi/abs/10.1080/1573062X.2024.2313634

[28] Bhardwaj P, Tiwari P, Olejar K, Parr W, Kulasiri D. A machine learning application in wine quality prediction. Machine Learning with Applications. 2022;**8**:100261

[29] Tyagi A, Tiwari P, Bhardwaj P, Chawla H. Prognosis of sexual dimorphism with unfused hyoid bone: Artificial intelligence informed decision making with discriminant analysis. Science and Justice. 2021;**61**(6):789-796

[30] Tiwari P, Bhardwaj P, Keprate A, Tyagi A. Breast cancer survival prediction using machine learning. Studies in Computational Intelligence. 2022;**1016**:143-158

[31] Zhu M, Wang J, Yang X, Zhang Y, Zhang L, Ren H, et al. A review of the application of machine learning in water quality evaluation. Eco-Environment and Health. 2022;**1**(2):107. Available from: https://pmc.ncbi.nlm.nih.gov/articles/PMC10702893/

[32] Tiwari P, Bhardwaj P, Somin S, Parr WV, Harrison R, Kulasiri D. Understanding quality of pinot noir wine: Can modelling and machine learning pave the way? Food. 2022;**11**(19):3072. Available from: https://www.mdpi.com/2304-8158/11/19/3072/htm

[33] Brooks W, Corsi S, Fienen M, Carvin R. Predicting recreational water quality advisories: A comparison of statistical methods. Environmental Modelling and Software. 2016;**76**:81-94

[34] Vidal V, Sampognaro L, de León F, Kruk C, Perera G, Crisci C, et al. A critical review of model construction and performance for nowcast systems for faecal contamination in recreational beaches. Science of the Total Environment. 2024;**954**:176233. Available from: https://linkinghub.elsevier.com/retrieve/pii/S0048969724063897

[35] Olawade DB, Wada OZ, Ige AO, Egbewole BI, Olojo A, Oladapo BI. Artificial intelligence in environmental monitoring: Advancements, challenges, and future directions. Hygiene and Environmental Health Advances. 2024;**12**:100114

[36] Shams MY, Elshewey AM, El-kenawy ESM, Ibrahim A, Talaat FM, Tarek Z. Water quality prediction using machine learning models based on grid search method. Multimedia Tools and Applications. 2024;**83**(12):35307-35334. Available from: https://link.springer.com/article/10.1007/s11042-023-16737-4

[37] Farrell ML, Joyce A, Duane S, Fitzhenry K, Hooban B, Burke LP, et al. Evaluating the potential for exposure to organisms of public health concern in naturally occurring bathing waters in Europe: A scoping review. Water Research. 2021;**206**:117711

[38] Russo GS, Eftim SE, Goldstone AE, Dufour AP, Nappier SP, Wade TJ. Evaluating health risks associated with exposure to ambient surface waters during recreational activities: A systematic review and

meta-analysis. Water Research. 2020;**176**:115729. Available from: https://pmc.ncbi.nlm.nih.gov/articles/PMC10287035/

[39] Smith DG, Maasdam R. New Zealand's national river water quality network, design and physico-chemical characterisation. New Zealand Journal of Marine and Freshwater Research. 1994;**28**(1):19-35

[40] Tait A, Henderson R, Turner R, Zheng XG. Thin plate smoothing spline interpolation of daily rainfall for New Zealand using a climatological rainfall surface. International Journal of Climatology. 2006;**26**:2097-2115

[41] Snelder TH, Biggs BJF, Woods RA. Improved eco-hydrological classification of rivers. River Research and Applications. 2005;**21**:609-628. DOI: 10.1002/rra.826

[42] Leathwick JR, West D, Chadderton L, Gerbeaux P, Kelly D, Robertson H. Freshwater Ecosystems of New Zealand (FENZ) Geodatabase: Version One – August 2010 – User Guide. New Zealand: Department of Conservation; 2010. pp. 1-51